The archives of the peat bogs

The archives of the peat bogs

SIR HARRY GODWIN, FRS

EMERITUS PROFESSOR OF BOTANY
UNIVERSITY OF CAMBRIDGE

CAMBRIDGE UNIVERSITY PRESS
CAMBRIDGE
LONDON · NEW YORK · NEW ROCHELLE
MELBOURNE · SYDNEY

Published by the Press Syndicate of the University of Cambridge
The Pitt Building, Trumpington Street, Cambridge CB2 1RP
32 East 57th Street, New York, NY 10022, USA
296 Beaconsfield Parade, Middle Park, Melbourne 3206, Australia

First published 1981

Printed in Great Britain at The Pitman Press, Bath

British Library Cataloguing in Publication Data
Godwin, *Sir* Harry
The archives of the peat bogs
1. Peat
2. Peat-bogs
3. Peat soils
I. Title
574.5'2632   S598   80-41495
ISBN 0 521 23784 X

Contents

Acknowledgments

Since this book concerns studies through a quarter of a century, it will be understood that my indebtedness is so great, and embraces so many colleagues and good friends, that it can only be expressed in general, though still sincere terms. Many of those most immediately concerned have been mentioned in the text: I trust that the rest will not take it amiss that I can only say collectively how warmly I recognise their kindness to and cooperation with me.

In building up familiarity with the techniques of bog enquiry, peat stratigraphy, pollen analysis and radiocarbon dating, one was able to draw freely upon the experience of scientists of many nations, especially those of western Europe and North America, and in the facilitation of meetings and working parties with them, Clare College and the University of Cambridge were extremely helpful. Similarly, throughout the founding and expansion of all the studies associated with Quaternary research the Research Councils, the Royal Society, the Nuffield Foundation, the British Council and the Wenner-Gren Foundation provided invaluable support.

A great deal of the investigation dealt with in the book relates to the Somerset Levels and I gladly acknowledge the immediate and effective facilities given by both individual peat diggers and by the larger firms engaged in peat extraction. The separate publications listed in the References have happily already allowed me to thank most of them individually, along with an extended succession of research students and friends who permitted themselves to be recruited into peat excavation, sampling, note-taking and haulage, lending not only brawn but acute and valued observation and criticism. In every way first of these was Professor A. R. Clapham, jointly with whom the first scientific publications of the derelict Somerset bogs were made. Material collected in the field has been analysed in the laboratory and I am deeply sensible of the sustained and skilful labour of the numerous laboratory staff who carried through this work, be it pollen analysis, plant identification or radiocarbon assay.

Now, at a later stage altogether, I am happy to acknowledge help in preparation of *this* volume. I have been able to make use of photographs by those consummate plant portraitists, Dr M. C. F. Proctor and Mr W. H. Palmer, by Dr Neville Moar and Mr K. G. Richman. The majority of photographs are my own but they owe much to their skilful printing and enlargment by Mrs Sylvia Dalton.

It is not difficult to appreciate the gratitude I feel to Professor Richard West, my successor both to headship of the Sub-department of Quaternary Research and to the Chair of Botany: working in the Botany School has made preparation of this book in many ways simpler and more efficient. Finally I allow myself the pleasure of reiterating the manifold thanks that I owe my wife, with whom the beginnings of pollen analysis were shared and who has jointly endured so many of the pleasures of bog-trotting and looking into the archives of the mires.

Some of the illustrations have previously appeared in scientific publications and I gratefully acknowledge their employment to the following journals, societies and book-publishers: *Acta Phytogeographica Suecica*, *Bristol Medico-Chirurgical Journal*, Clarendon Press Oxford, Kooperativa Förbundets Bokförlag-Stockholm, Nordisk Oldkundighed og Historie, Somersetshire Archaeological and Natural History Society, *Ulster Journal of Archaeology*, University of Bristol Spelaeological Society. Many of the author's own figures have been reproduced from the *Journal of Ecology, New Phytologist, Proceedings of the Prehistoric Society* and the journals of the Royal Society.

1

Introduction:
Quaternary research and mires

Those, who, like myself, get intense pleasure from visiting the solitary unspoiled stretches of our British countryside will recognise that few landscapes have been less affected by human interference than the water-logged peat bogs, especially the acidic mires so poor in mineral nutrients and so badly drained that they offer little inducement to cultivation, pasture or afforestation. Remote and guarded by legends of unfathomable morasses and malign spirits, they have retained a loneliness which is a welcome safeguard to the wild creatures native of them. To those of us anxious to discover the quality of the natural landscape and the behaviour of its natural plant and animal communities, these acidic mires correspondingly attract our deep interest, and under wide skies where one hears nothing beyond the call of the curlew down the wind, one comes to regard them with instinctive affection.

I have little doubt that such reactions played their part in persuading me in 1935, whilst continuing research in the East Anglian Fenland, to extend both ecological and stratigraphical research to the acidic bog lands of the north and west. Thus began a delightful involvement with what might light-heartedly be called bog-trotting, although heaven knows, there can be few natural communities less adapted to sustain trotting than those of the squelchy rain-fed mires.

Already by 1935, five years of field and laboratory research, done in conjunction with the Fenland Research Committee, had made it apparent that there now presented itself the chance to enter an almost virgin field of study. This was to become known as 'Quaternary research', a title invented to cover the very wide range of scientific and historical interest involved in the study of events and processes of the Quaternary Epoch, that is, the geological period covering the waxing and waning of the great glaciations, the mild intervening periods and the intensely important Flandrian period that has succeeded the latest of the major glacial stages.

The events leading to this conviction I have already described in *Fenland: Its Ancient Past and Uncertain Future* (1978). What I hope now to recount is how the switch to the great western peat bogs as a central area of research provided such stimulus and reinforcement to my conviction that the conclusions derived in the following twenty-five years or so became very closely bound up with the story of Quaternary study as a whole. Accordingly, to recount what these years reproduced is possibly to serve a more general historical purpose, and especially so if one can recapture the excitement and pleasure of the day by day discovery and progress of our work.

In a memorandum to the University of Cambridge in 1938 I set out the case for its establishing a research organisation for the encouragement of coordinated research from diverse scientific angles, into the events of the Glacial and Post-glacial periods. The war having naturally halted any such development, the case was restated in 1943, and eventually led to the creation in 1948 of a 'Sub-department of Quaternary Research' administratively tied to the Faculty Boards of Biology 'A', Geography, Geology, and Archaeology and Anthropology. I was made Director of it, a post I held until 1966. The Sub-department has accordingly now enjoyed upwards of thirty years of activity, during which time, not only here within Cambridge precincts, but throughout the world, Quaternary research has attracted enormous popularity and successful development, and many scientists of great distinction are engaged upon it and constantly are achieving fresh successes in exploring its possibilities.

These circumstances make it essential that I declare at once that this book is no more than an account of my own purely personal involvement with Quaternary studies and their ecological background, and that it barely extends beyond 1960, although the time-space 1935–60 enables me to attempt some reconstruction of the environment in which the young subject of Quaternary study was being first defined and applied. The year 1955 could no doubt be regarded as of critical significance for the young Sub-department. Dr Donald Walker (later to become Professor in Canberra) had returned from National Service to take up his post of Senior Assistant in Research, Dr Richard West (later to be head of the Sub-department and Professor of Botany in Cambridge) had just taken the more junior post of Assistant in Research, and Dr Eric Willis had designed, built and brought into commission our own radiocarbon dating laboratory. They and numerous research workers of less permanent grade comprised an extremely active centre of expanding advance, as is very well shewn in the successive reports published each year in the *University Reporter*.

Finally by 1956 I had completed and published the comprehensive *History of the British Flora* that gave an organised presentation of all the detailed evidence made available by the methods of Quaternary research, and gave a summary of what those methods were and the main conclusions to which we had been led. It was not long before election to the Chair of Botany (1960) and the Presidency of the Xth International Botanical Congress (1964) were to provide such considerable involvement with administration and paper-work that personal concern with field or even laboratory work dwindled to a small, though persistent stream.

All the same, to limit oneself to this early phase of Quaternary research has its advantages. The pioneer is concerned, as at no later time, with the most general and basic conceptions: he is not embarrassed by preconceived notions, but has (in the wise words of a great American inventor) 'the advantage of beginning from total ignorance'. It is also certainly true, as Lord Zuckerman writes concerning his war-time research, that in most fields of scientific enquiry the cream of the new intellectual adventures is skimmed off in the first year or two after starting! We were a little slower. It is especially of such early phases of research that one can say 'a good hypothesis is worth a ton of facts', whilst accepting that of course a good hypothesis is no alternative to the organising of established proof, but a great stimulus to the effective collection and testing of relevant information. The discoveries and implications of this early phase of Quaternary research, in Cambridge and elsewhere, certainly caused all kinds of further enquiry to be started, much of course by one's own students following promising leads and applying generalisations from one area to another. Such enquiries were part of a wave of enthusiastic development of Quaternary study everywhere, but it would be quite contrary to my present intention to attempt any presentation of this mass of active research or to try to evaluate the present state of knowledge in this vast scientific field. Only in completion of important arguments already developed have relatively recent results been mentioned, as in Chapters 12 to 14.

In the first years of my botanical career, even then a convinced ecologist, I took advantage of the nearness to Cambridge of a widely known nature reserve, Wicken Fen, to engage upon a series of field studies, observations and experiments. In these I sought to make out the factors relating the major plant communities to one another, and especially I tried to discern the natural processes by which, as peat accumulates in conditions of maintained water-logging, a consistent process of vegetational succession inevitably takes place.

This sequence, the so-called hydrosere, begins with the slow deposition of organic detritus in the open water of flooded valleys or basins. Progressive shallowing in time allows invasion by submerged water-

Plate 1. Fen vegetation at Wicken, Cambridgeshire. The calcareous waters support lilies and marginally reeds and reed-mace. The emergent alkaline peat supports dense bush growth, especially of sallow and buckthorn, that later develops into fen-wood with ash and oak.

plants (such as the pond-weeds and water-lilies) upon the bottom mud and subsequently by the tall erect stems of the reed-swamp dominated by bulrush (*Scirpus lacustris*), common reed (*Phragmites communis*), various small reeds and sedges, and particularly the giant sword sedge (*Cladium mariscus*). Peat growth, speeded up by the addition of the underground parts of these vigorous plants, now allows the establishment of such shrubs as the sallow (*Salix cinerea*) and hairy birch (*Betula pubescens*), and later of alder (*Alnus glutinosa*), with even oak (*Quercus robur*) and ash (*Fraxinus excelsior*), eventually to form fen-woods. We established the general nature of these successional (seral) relationships and were able to show how human cropping of the vegetation had often modified it.

At Wicken Fen we were concerned with a category of peat land and peat land vegetation that can be recognised very often in low-lying, constantly flooded areas. This main type of peat land (now conveniently referred to as 'mire' in conformity with the Swedish usage 'myr') is known to all English-speaking ecologists as 'fen'. Its first essential quality is that it has a *topogenous* (i.e. landscape dependent) origin. The water-logging responsible for its occurrence is induced by the convergence in one limited valley or basin of the surface drainage of a bigger area. Thus in the East Anglian Fenland the great river systems of Witham, Welland, Nene and Great Ouse (with its large westward-running tributaries) all discharge into the wide lowlands that surround

The Wash: here drainage is hindered by the building up of tidal silts at the seaward side, ponding back the abundant inflow from the rivers. It was in such conditions that deep peat (still 10 ft (3 m) or more at Wicken) was able to form throughout the 'Black' Fens. The second essential 'fen' quality arises from the constant importation with the river water of dissolved mineral salts that provide high concentrations of inorganic nutrients, important alike to the submerged plants of the early seral stages and to all those that follow. The conditions and the plant communities are in consequence spoken of as *eutrophic* (i.e. of good nourishment). Even where the drained landscape is one of thin soils and acidic rocks this *eutrophication* is recognisable (although to a minor extent) in the increased variety and richness of the flora in the drainage basins, an effect that is also passed on, naturally enough, to the quality and variety of animal life dependent for nutrition on the plants. The streams, however, that drain into the East Anglian Fenland traverse a landscape of quite another kind, made of soft sedimentary rock of relatively recent geological age, including big stretches of the Chalk and substantial exposures of Jurassic limestone. Moreover, the several advances of ice in past glaciations have scraped and eroded exposures of these formations, leaving the ground mantled everywhere with a highly calcareous till. Thus the river waters have been fully charged with soluble calcium bicarbonate and the accumulating peat throughout its depth has been made alkaline in reaction. So long as the flat fen surface remains within reach of the calcareous flood-water, the natural tendency of decaying vegetative matter to become acidic upon oxidation is fully compensated. Finally, as is always the case, the vegetation, by its response to the environmental conditions, itself directly indicates the specific fen type. It is dominated in all the middle hydroseral stages by a great variety of monocotyledons (sedges, grasses and rushes in a great range of species and often of large size), with equally vigorous dicotyledonous plants able to co-exist with the monocots (such for example as *Angelica*, hemp-agrimony (*Eupatorium cannabinum*), purple loosestrife (*Lythrum salicaria*), yellow loosestrife (*Lysimachia vulgaris*), milk-parsley (*Peucedanum palustre*), marsh thistle (*Cirsium palustre*), and very many more). In the later hydroseral stages, as I have said, woody plants right up to the stage of mature fen-wood are entirely typical of this mire type. Where the trophic (nutritional) condition of the drainage water is less extreme, as over the base-poor hard igneous rocks, a mire type that may be called 'acid fen' can develop, but for such development we need to look to the mountains of the north and west of this country.

We may note, *en passant*, how its hydrographic nature makes the fen mire to a very large extent independent of climate, for even in regions of

Plate 2. Fen-wood in valley of River Beaulieu, Hampshire. The many-stemmed alders have largely killed out the herbaceous vegetation of the preceding fen stage and the mattress it formed is breaking up.

low rainfall, local water-logging is induced essentially as a consequence of the local topography and local concentration of the stream and river drainage.

The second great group of peat lands (mires) is that of the bogs or mosses. Although these English colloquial terms have a wide application and the term 'moss' extends of course to the lowly cryptogamic plants, the Bryales, nevertheless in their central areas of usage the terms coincide well enough with the category of ombrogenous mires, that is to say, those dependent for sustained water-supply merely upon direct precipitation either as rain or snow. Such mires will necessarily be dependent upon climate, being favoured on the one hand by high precipitation and on the other by a low evaporation rate, in short by increased oceanicity, and it is no accident that they occur most abundantly in the atlantic west of the British Isles, as in north-western Europe generally. Since the water-supply comes entirely from the sky it is naturally poor in plant nutrients, receiving only air-borne dust, traces of salt derived by long-distance carriage from sea-spray, and traces of ammonium washed from the air. It makes a poor diet for an extensive plant carpet, and certainly such mires and their plant communities are qualified to be classed as *oligotrophic* (of meagre nourishment). A portmanteau term sometimes used is *ombrotrophic*: the English equivalent – rain-fed – seems specially evocative. In such mires the growing bog surface with its acidic products of humification lacks

Fig. 1. Diagram to shew idealised raised bog occupying a flat valley floor. When drainage from the bog meets that of the hillside a wet lagg is formed that gets wider downstream: because it is richer in mineral bases its vegetation is then more fen-like and it inhibits invasion by the *Sphagna* of the bog. Thus the bog margin (rand) is highest and steepest downstream and least pronounced upstream where almost no lagg separates the bog from the gently sloping hillside. (After Osvald.)

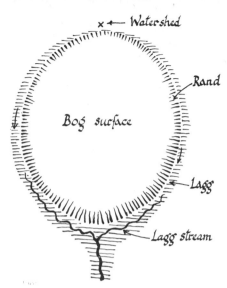

virtually all neutralisation by bases, so that the peat is acidic throughout. Thus in all ombrogenous mires we find ourselves contemplating plant communities dominated by the same highly characteristic range of species that are oxylophytes (of acidic soils), even possibly calcifuges (lime-hating), that tolerate base-poor, water-logged, or at least very wet conditions and correspondingly often favour moist climates. Of this assemblage the key genus is the bog moss *Sphagnum*, whose many species occupy various ecological niches in the vegetation of these acidic mires, many of them indeed playing a primary role in building up the bog peat. They occupy this central role to a large extent by virtue of their remarkable powers of water-retention within large empty porous cells that constitute a large proportion by bulk of their leaves and stems, and that convert the adjacent shoots of the *Sphagnum* hummock into an extremely effective sponge, capable of sustaining the plant colony through dry periods of considerable length. Other extremely characteristic oxylophytes of the bog surface are the various members of the calcifuge heath family, the heathers (*Erica tetralix* and *E. cinerea*), ling (*Calluna vulgaris*), cranberry (*Vaccinium oxycoccus*), bilberry (*V. myrtillus*) and *Andromeda polifolia*, a plant that has no English name in common usage. Also, from other families, we have the crowberry (*Empetrum nigrum*) and cloudberry (*Rubus chamaemorus*). Such species, mostly shrubby, tend to occupy the higher areas of the bog surface along with sheathed cotton-grass (*Eriophorum vaginatum*), deer grass (*Scirpus caespitosus*) and purple moor grass (*Molinia caerulea*), whilst in the wetter hollows are the beaked sedges (*Rhynchospora alba* and *R. fusca*), bog-

Plate 3. Raised bog in County Athlone, Ireland, shewing the natural gentle dome of the uncut central surface of the mire. (1935)

asphodel (*Narthecium ossifragum*), the many-headed cotton-grass (*Eriophorum angustifolium*), and, more marginally, the three British sundews (*Drosera rotundifolia, D. anglica* and *D. intermedia*). How strongly such a plant assemblage contrasts with that which characterises the eutrophic fens, and how forcibly this contrast makes the point that the plant communities themselves, responding individually and collectively, are the best ultimate indicators of the prevalent habitat conditions and thus also of the major mire types! This conclusion would be only strengthened were we to extend mention to include in our consideration the great wealth of mosses and liverworts that also characterise, at ground-level, the living carpet of the bogs.

The ombrotrophic mires are divided into two major groups, the raised bog and the blanket bog, part of the terminology, along with 'mire' itself, hammered out by Professor A. G. Tansley, Hugo Osvald and myself during our meetings in 1935, and subsequently brought into general use. The raised moss or raised bog develops in the less extreme conditions of oceanicity and takes the form of a gently domed cushion of peat, often of considerable lateral extent (very often a mile (1.6 km) or more across) and a height of several feet (up to 5 m or more) above the immediately surrounding landscape. It is the equivalent of the German *Hochmoor* and the Swedish *Högmosse*, both terms indicative of the shape of the mire. The centre of the raised bog is very nearly flat, so that the attempt to popularise the architectural term 'cupola' for the bog centre is, I think, ill-advised. The margins of the raised mass of peat slope downwards relatively steeply and are cut by little drainage streams from the bog surface. These, at the foot of the 'rand', as the sloping margin is known, meet the collected drainage from the adjacent mineral soil to form a slightly eutrophic fen known as the 'lagg', which on the downslope side may include a small stream that deepens progressively in its lower reaches. It is the base content of this lagg drainage that inhibits lateral growth of the *Sphagnum* peat, so that where the lagg is deep and more eutrophic the rand of the bog remains high and steep: the acid mire develops most readily upstream where the lagg is at its feeblest (Fig. 1).

In considering the ombrogeneity of the raised bog I like to contemplate the notion that if a billiard table were left out of doors in a suitable climate, such as that of central Ireland, lowland Cumbria or Wales, it would, if long neglected, grow a raised bog upon its level surface! I have in fact observed something akin to this in County Sligo, where a tall stack of Mountain Limestone stood detached from the main escarpment: its flat surface, only a few square metres in extent, was altogether prevented from receiving any water-supply save that from the sky and yet it was capped by a considerable depth of acidic peat and carried a typical bog flora. No doubt to some degree the surface of the limestone had been sealed by the residues of solution and the peat was unaffected by the chemical properties of the underlying rock. This Sligo site is in fact on the atlantic side of the climatic boundary, possibly around the 40 in (100 cm) annual rainfall line, that is taken to mark the transition from the region of raised bog to that of blanket bog. As the rainfall, or better the rainfall–evaporation ratio, increases, it appears that direct

Plate 4. Blangsmosse, central Sweden, shewing the edge of an unspoiled raised bog. Left, the sloping rand of the bog covered in dwarf shrubs, small pine and birch; right, coniferous woodland (spruce) on rising mineral soil; centrally, the flat lagg kept wet and fenny by drainage water from both sides and dominated by sedges. (1958)

Plate 5. The perfect demonstration of an ombrogenous (rain-fed) mire, Carrowkeel, County Sligo, western Ireland. On the flat top of this perched limestone block there is a considerable depth of acidic peat that supports such lime-intolerant plants as *Calluna* (ling). The weathering of the limestone is advanced and the block must have been long detached: the atmosphere alone must be source for the water that sustains peat growth on this island bog. Lough Arrow and drumlin landscape in background. (1949)

Plate 6. Escarpment of the Carboniferous Limestone at Carrowkeel, County Sligo, with acidic peat bog sitting directly upon the limestone. The bog margin sustains abundant tall bushes of *Calluna* and the peat face shows some buried timber. The very high annual rainfall of this region (over 1000 mm) naturally conduces to ombrogenous bog formation. (1949)

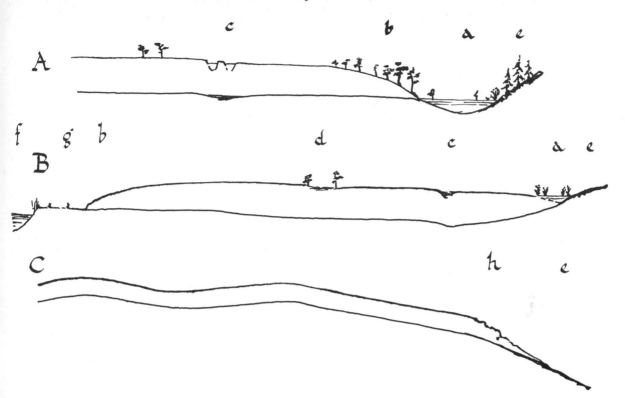

Fig. 2. Morphology of the two chief types of ombrogenous (rain-fed) mire. *A* and *B* are both raised bog; *A*, the more wooded one, that found commonly in sub-continental climates (e.g. south Sweden), *B*, unwooded, that typical of the British region. *C*, blanket bog: *a*, lagg; *b*, rand; *c*, bog pools; *d*, 'soak' on bog surface with fringing trees; *e*, adjacent mineral soil; *f*, river limiting spread of raised bog, with *g*, river flood-terrace; *h*, eroding margin of blanket bog at increased hill-slope.

precipitation not only suffices to maintain water-logging upon flat surfaces, but that it will do so even upon gentle slopes. In terms of my visual image, one would still get bog growing on the billiard table even if it were tilted up at one end! Thus in the atlantic climate of western Ireland one may see a continuous cover of acidic peat, 3 to 6 ft (1 or 2 m) in thickness, extending up slopes as great as 10 or 15 degrees. Since flat summits, saddles or basins are likewise peat-covered, an almost continuous mantle of mire stretches across the landscape. This is the mire type known in Swedish as *Täckmosse*, in German as *Terrainbedeckender Moor*, and in English as blanket bog or blanket moss. When the hill-slopes are thus blanketed, the drainage water moving downhill is inevitably very base-deficient, so that basins thus fed are themselves acidic and only slightly eutrophic: they are indeed 'acid fens'. A similar consideration applies to those raised bogs growing on the atlantic side of their distribution range: the lagg is so acidic and oligotrophic that it offers little obstacle to the lateral spread of the bog mosses, the rand is flat or non-existent at the contact with the hillside, so that the raised bog takes on the transitional (and anomalous) character of 'flat raised bog'. By contrast, at the dry side of their climatic range the raised bogs develop very strongly sloping margins and respond to the relative dryness by carrying abundant trees, mostly of pine. Beyond this climatic limit there is no development of ombrogenous bog: the mires are all topogenous with greater or less development of the character of fens, but at which precise climatic level the transition occurs is not agreed; it could well be near the 20 in (51 cm) isohyet.

A botanist who had hitherto very largely confined his ecological investigations to fens of the most characteristic and eutrophic kind, had to realise how the acidic bogs strongly contrasted with his experience, and their study offered acquaintance with quite different habitat types and indeed with a flora and communities seldom encountered in the dry eastern part of Britain. There was also the fact that studies in the stratigraphy of the peat deposits of the Fenland basin had already indicated that this region had perhaps not entirely been without its own raised bogs. In limited areas less affected by the flooding of lime-rich river water, it appeared that in the past there had developed acidic peat with *Sphagnum*, cotton-grass and ling. There were indeed residual herbarium and written records indicative of the past growth of similar calcifuge plants. This evidence was subsequently to be multiplied and confirmed, but even by 1935 it offered considerable inducement to making oneself more familiar with the ombrogenous bogs where they grew in their still-active state.

An equally powerful inducement to interest oneself in the ombrogenous bogs, more especially the raised bogs, arose from the fact that my wife and I had, since 1931, become very thoroughly involved with the practice of pollen analysis, the research method that had been developed by the Swedes G. Lagerheim and L. von Post, and popularised in the English-speaking world by G. E. Erdtman, who had in fact applied it in a preliminary way to chosen deposits in several parts of Britain. It is not necessary now to describe the method except to say that it is essentially a means of recording past vegetational history through the identification of sub-fossil pollen grains preserved in water-logged deposits such as lake muds and peat. Pollen is produced in such vast amount and shed so freely in the wind that receptive surfaces no larger than one's fingernail will receive many thousands of pollen grains each season. Add to this the fact that in anaerobic conditions the pollen grain membranes have most remarkable powers of resistance to decay and that they have a wonderful range of shapes, sizes and sculpturing visible under the microscope, and it will be seen what rich potentialities their study will offer. A statistical count of the proportions of all the major pollen types falling on a bog surface will reflect, albeit with some understandable distortion, the main vegetational components of the surrounding vegetation. Correspondingly a sequence of

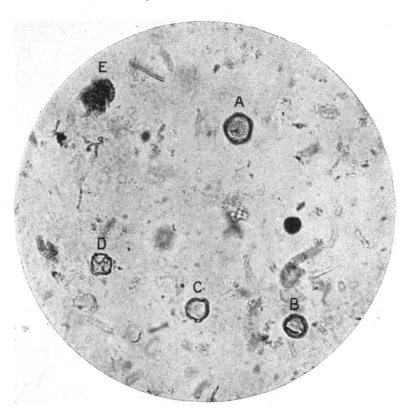

Plate 7. Microscopic preparation of a lake mud shewing among the detritus (such as E) tree-pollen of A, *Ulmus* (elm); B, *Quercus* (oak); C, *Betula* (birch); D, *Alnus* (alder). (1941)

samples from a deposit that has formed through a long period of time will also, upon analysis of its fossil pollen, reflect the changes in vegetation that have taken place throughout formation of the deposit. It is in fact surprisingly easy to separate the fossil pollen from the enclosing matrix because the pollen membranes have proved extremely resistant to the powerful chemical reagents that are used to remove clay, silt and the abundant humic colloids and cellulosic plant remains that enclose them. Already by 1930 it had become clear that the long records of changing pollen composition obtained by analysing vertical series through deep peat deposits shewed remarkable similarities in the vegetational history they displayed, a history extending right from the time of the last glacial retreat up to the present day. It was evident that in south Sweden, for example, treeless vegetation had been successively invaded by birch forest, then by pine, and subsequently by the more warmth-demanding trees of the deciduous forest, pioneered by the hazel in large amount and composed of elm, oak, alder and to a lesser degree by lime. Later still the sequence shewed the spread of spruce, and to some extent also of beech and hornbeam. At quite an early stage in our own investigations it became clear that a similar pattern of forest history could be made out in East Anglia (Fig. 3). As time has gone on, diagrams from all parts of the country have been shown to exhibit conformable changes, as will be seen by examples given later in the book. There is little doubt that these changes in woodland composition, through the prehistoric millennia when forest was the natural unbroken 'climax' vegetation, were directly under climatic control, so that the correspondence of pollen diagrams over large regions was not unexpected. Given some care in interpretation, the successive stages of forest dominance might thus be utilised as a kind of coarse chronological scale, a usage made all the more attractive since the pollen grains were so unbelievably abundant that they could be recovered from an immense range of situations, and yielded useful results from samples as small as a wheat grain. In the days long before radiocarbon age determinations the attractiveness of a ubiquitous chronological scale of this kind was very great. It was a scale moreover that with diligence could be brought into correlation with many other indices of change through the last 10 000 years of climatic amelioration since the Ice Age. Thus the forest history of southern Scandinavia was soon brought into relation with the successive post-glacial climatic periods suggested by the Norwegian A. Blytt and the Swede R. Sernander.

Fig. 3. A generalised pollen diagram for East Anglia to shew the changes in relative abundance of the various types of tree-pollen through the whole Post-glacial (Flandrian) period. The earliest dominants are cold-tolerant (birch (*Betula*) and pine (*Pinus*)) but they give way to warmth-demanding genera of the mixed-oak forest as the climate reaches its optimum. Beech (*Fagus*) and hornbeam (*Carpinus*) are characteristically late in their appearance.

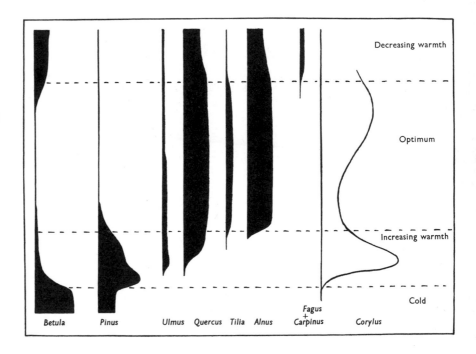

Early chronological scale: southern Scandinavia, 1931

Climatic period	Character	Forests
Sub-atlantic	Cool and wet	Spreading beech, spruce
	500 B.C.	
Sub-boreal	Dry, warm and continental	Mixed oak forests
	3000 B.C.	
Atlantic	Warm, moist and oceanic	Mixed oak forests
	5600 B.C.	
Boreal	Dry, warm and continental	Hazel scrub, pine
	7500 B.C.	
Pre-boreal	Cool summers, severe winters	Birch
	8000 B.C.	

The forest stages were tied into changes of relative land- and sea-level and the related phases of fresh and salt water in the Baltic. They were even hitched on to a provisional absolute time-scale based upon analysis of layers of sediment annually laid down in the lakes fronting the retreating Scandinavian glaciers. A special appeal, however, came in the correlation of the forest history chronology with the stages of prehistoric archaeology, previously datable only by rather devious and chancy comparisons extending back to the time-scale of ancient Egypt. Raised bogs in particular had proved very suitable repositories for the interred remains of prehistoric man, where, although the acidic medium dissolved the bony skeleton, proteins were well preserved and actually tanned by the bog water, so that human bodies from as long ago as the Iron Age had skin, hair and stomach contents fully preserved. Wooden objects, tools, weapons and buildings were likewise almost intact so long as water-logging had been continuous, and kegs of bog-butter, palatable on excavation to the local dogs, were familiar to the Irish peat diggers. Objects such as these, when seen *in situ* in the bog, can often be placed accurately within a pollen series from samples on the spot, and may thus be referred safely to a stage within the pollen zonation. In contrast with the loose unconsolidated muds of a lake bottom, the fibrous nature of peat tends to preserve the original stratigraphic relation of discarded artefacts, as we had ourselves seen in the East Anglian Fenland, but the acid mires offered one special advantage over the fens. In the latter, there is always a very heavy influence of local factors upon the pollen diagrams, especially when the later stages of the hydrosere are represented in the peat by fen-woods, that induce such a preponderance of local birch and alder pollen as to obscure the general, climatically governed, drift of the region as a whole. The ombrotrophic

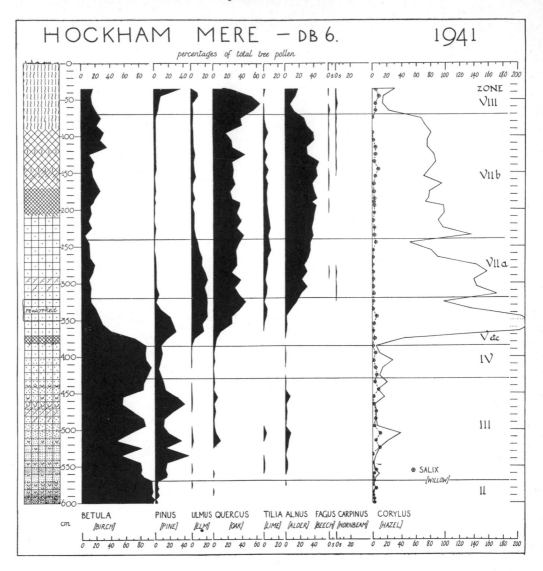

HOCKHAM MERE — DB 6.   1941

percentages of total tree pollen

BETULA [BIRCH] PINUS [PINE] ULMUS [ELM] QUERCUS [OAK] TILIA [LIME] ALNUS [ALDER] FAGUS [BEECH] CARPINUS [HORNBEAM] CORYLUS [HAZEL]

⊕ SALIX [WILLOW]

bogs, in the British Isles at least, carry trees only sparsely and the pollen diagram from the centre of a large raised bog can be trusted to reveal the forest history of the whole surrounding region. Accordingly, when in the early 1930s we were seeking to compile a network of long pollen diagrams across the whole country, beginning with England and Wales, we welcomed the likelihood that in the north and west we might find suitable records within the raised bogs, whether intact or already partially cut away.

The opportunity to realise some of the potentialities of investigating the ombrogenous peat bogs, ecologically and historically, first presented itself in 1935 and the following chapters tell the story of our involvement through the ensuing twenty-five years. In particular they are intended to display the role of such mires as repositories of post-glacial history extending through several millennia: our hope is that they will make an effective and even entertaining introduction to this very fertile and still expanding field of palaeoecology.

This book has not the slightest claim to be a text-book or critical review of the considerable volume of scientific work published on British peat bogs from 1960 onwards. The broad brush strokes of the conclusions we discerned in the years before 1960 have been followed by a detailed and abundant delineation of processes, structure and history in mires of all types and from most parts of the British Isles. It will be for some younger, active participant to evaluate the product of these latest twenty years. We shall be satisfied for our part to have set down the common basis, historical and theoretical, from which the newer knowledge has arisen, and to have exhibited the potentialities of translating these unlikely sources into such wealth of information about the past.

◀

Fig. 4. One of the earliest pollen diagrams from East Anglia to embrace practically the whole Flandrian (Post-glacial) period. All except the uppermost metre consists of lake muds and the pollen curves are remarkably free from local distortion. The proportion of each tree-pollen type is expressed as a percentage of the total tree-pollen (as are hazel and willow). The pollen zonation is that employed for England and Wales.

2

Living raised bog

Ireland, 1935

By 1935 Professor A. G. Tansley had sufficiently achieved the reorganisation of the Oxford Chair of Botany to feel himself free to begin the rewriting in expanded form of a small, but famous book he had published in 1911 as *Types of British Vegetation*, and as part of the preparatory study he arranged a botanical excursion to Ireland to examine the natural bog communities still to be seen there in fairly undisturbed condition. He had secured the collaboration of Hugo Osvald, then just retiring as Director of the Swedish Peat Reclamation Society's Institute at Jönköping, author of the great classic monograph on the extensive Swedish raised bog system of Komosse, and one of the world's greatest authorities upon peat bogs, their communities, structure and utilisation. It was a stroke of infinite luck that I was invited to complete the party and I gratefully accepted the role of chauffeur of the old 'T' model Ford tourer that we hired in Dublin in which to rattle (successfully) over the countryside. Both A.G. and I drew unreservedly upon Hugo's vast experience and by that best of all methods, direct demonstration in the field, we were shewn most of the major features of Irish bog ecology and became familiar with his own system of bog classification.

My ignorance of bog ecology was vast, and I was immensely impressed by all I saw, most of all by the great raised bogs that then occupied a large part of the flat central Irish plain, collectively known as the Bog of Allen. One was immediately struck by the vast extent of the undrained bog, with its almost treeless surface extending a mile or more in every direction with no recognisable change in slope. The bog surface very commonly carried a complex mosaic of open shallow pools and low hummocks of red *Sphagnum* moss, and I recall how timorously I initially made my way across it. In fact there is little need to fear being 'bogged down': only the soft unvegetated pools fail to bear one's weight, and the ling, heather, cotton-grass and deer grass of the tussocks easily support

Plate 8. Regeneration complex on a Swedish raised bog shewing a pool with submerged *Sphagna*, invaded by rapid growth of the red hummock-building *S. magellanicum* (right centre). Deer grass and cotton-grass are abundant tussock-formers also.

one. Only here and there does one encounter the wide shallow band of a surface-drainage channel or bog 'soak', where there are no firm tussocks but an open mat of floating *Sphagna*, the loose rhizomes of bog-bean (*Menyanthes trifoliata*) and the many-headed cotton-grass (*Eriophorum angustifolium*), with a scatter of soft rush (*Juncus effusus*) and various sedges. The jay-walking botanist who fails to notice the change in texture and composition of the plant cover will quickly find himself in deep water.

Here on the squelching bog surface of pool and hummock Osvald demonstrated convincingly how the spatial distribution of plants conformed to the small-scale topography and how both were intimately involved in the process of bog growth that he, following his famous teacher, Rutger Sernander, regarded as the 'regeneration cycle'. What was involved can be illustrated by the sketch-diagram from my own field notes:

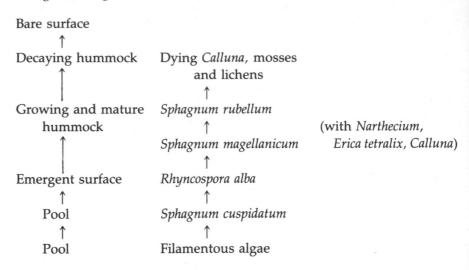

Bare surface

↑

Decaying hummock   Dying *Calluna*, mosses
                                              and lichens

↑   ↑

Growing and mature   *Sphagnum rubellum*
   hummock   (with *Narthecium,*
                                               ↑   *Erica tetralix, Calluna*)
                        Sphagnum magellanicum

↑   ↑

Emergent surface   *Rhyncospora alba*

↑   ↑

Pool   *Sphagnum cuspidatum*

↑   ↑

Pool   Filamentous algae

The shallow dark-brown pools, up to 3 ft (1 m) or so across, were bare of vegetation save for the tangled web·of algae and the submerged shoots of the aquatic *Sphagnum cuspidatum* whose remains could be seen to be filling up the pools so that marginally the emergent surface was colonised by other species, notably the hummock-building *Sphagna*. Here these were especially *Sphagnum magellanicum*, *S. rubellum* and *S. plumulosum*, all three of them so wonderfully crimson as to earn the Irish raised bogs the deserved name of 'red bogs'. As the compact branches of

Plate 9. The growing surface of raised bog shewing a small open pool now largely obliterated by a hummock of active *Sphagna*. The tufted plants are deer grass and cotton-grass; much cross-leaved heath is present on the older hummocks.

Plate *10*. Raised bog, shewing the centre of an old dying hummock; it carries large plants of ling (*Calluna*), but the centre is occupied by a carpet of grey lichens.

such mosses build the rounded hummock, conditions are made locally suitable for many of the species of higher plants so characteristic of acidic peat bog. At an early stage come in the beaked sedge (*Rhyncho-spora alba*), the cotton-grass (*Eriophorum angustifolium*) and common sundew (*Drosera rotundifolia*), whilst a little later, as the mound grows, we have bog-asphodel (*Narthecium ossifragum*) and the seedlings of plants like the ling (*Calluna vulgaris*), cross-leaved heath (*Erica tetralix*) and deer grass (*Scirpus caespitosus*), that accompany later stages of hummock growth and often finally dominate them. Finally one sees that in the centre of many large hummocks the *Sphagna* appear yellow and diseased, with abundant small-leaved liverworts embedded between their compact tufts of branches and with a variety of lichens appearing upon the dying *Calluna* twigs and on the broken peaty surface of the hummock itself. A sequence of such a kind as this, but with a great variation in the detail of the species concerned (a good deal more numerous than those here mentioned), occurs all over the bog surface, and we had no doubt it represented a natural successional process, an *autogenic* one, since it was induced by the reaction of the plants themselves upon the habitat. Osvald now proceeded to persuade us further by taking off his jacket and plunging his bare arm down through the bog surface, to bring up in his hand from successive depths the peat characteristic of different stages of the growth cycle: the shiny, struc-tureless mud of the open pool, the structured compact shoots of the mossy hummock builders, the rhizomes of bog-asphodel and cotton-grass, the stem-bases of deer grass and purple moor grass and finally the crooked broken woody stems of the ling in the residue of the decayed hummock surface. This sequence could be demonstrated to be repeated as far down as one could reach, and the series was easily extended by using a chamber-auger for sampling. When, at a later stage in our trip we inspected the peat cuttings that eat into the margins of the majority of the raised bogs, we were able to see exposed on the vertical peat faces the dramatic pattern of flattened lenses so typical of the pool-and-hummock peat. Here the body of each lens was made of the pale compact and undecayed shoots of one or other of the hummock-building *Sphagna*, seated upon slightly fibrous or horizontally layered greenish pool peat and overlaid by dark-brown humified *Calluna* peat (often associated with deer grass or cotton-grass tussocks) from the top of a mature or dying hummock.

All this evidence combined to indicate that the surface of the raised bog is growing upwards by the activity of the regeneration cycle. As a given hummock reaches its terminal stage and ceases growth it is being overtaken by the growth of infilling pools and rising hummocks alongside: indeed it helps their growth by shedding its own drainage into them. As the neighbouring hummocks now swiftly rise, the old hummock centre itself begins to be water-logged and before long lies below a new open pool. Thus the bog surface can be thought of as porridge on a slow boil, fresh bubbles (hummocks) constantly arising in one place after another and then subsiding. The cyclic growth is not in phase nor do the various parts of the surface necessarily complete the full cycle suggested by our diagram. Half-completed hummocks may be overtaken by flooding, and extensive and tall tussocks may persist

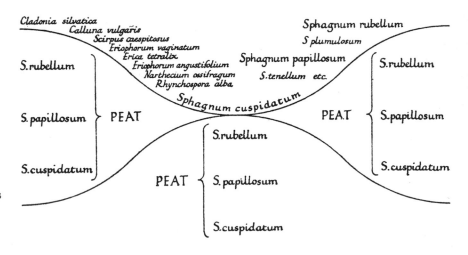

Fig. 5. Diagram to shew the succession of species in the process of the infilling of bog pools, the growth and decay of moss hummocks, i.e. the regeneration cycle. (After Tansley, 1939.)

Plate *11*. The steep margin (rand) of a large raised bog at Ballykilleen, near Edenderry, abutting on the flood-plain of a strongly calcareous river full of water-plants characteristic of fenlands, as is the grazed flood-terrace. The bog cannot invade, so that the rand is steep and well drained: it is therefore colonised by bracken and furze and will support large (planted) trees. (1935)

unusually long, but in these bogs at least, as in the great Swedish Komosse, the essential vegetational role of the 'regeneration complex', as Osvald named this vegetation type, seemed fully demonstrated.

The domed shape of the red bogs was unmistakable despite the irregular peat cuttings at the margins. Climbing by these rectangular faces one ascended as much as 15 or 20 ft (4.5 to 6 m) to the top of the bog; the original shape of the rand had been destroyed but the drainage had been further sharpened so that the heather and often bracken grew tall and dense upon it. It was interesting to find near Ballykilleen an untouched bog margin next to a considerable river whose waters were still fully calcareous with a rich aquatic vegetation entirely typical of fenland. Behind its banks, raised by dredgings, was a flood-plain some 90 ft (27.5 m) wide full of soft rush, but behind this rose steeply the intact rand of the bog, densely clad with tall bracken and gorse, with here and there small clumps of self-sown birch. This steepness resulted from the check upon lateral extension of the bog by the calcareous stream, and it was responsible for the characteristic rand vegetation, above which one could penetrate by progressively wetter zones to the pool-and-hummock complex of the flat bog plain. Only in one or two places of contact between the bog and the mineral soil, here representing part of an esker ridge, had there been preserved limited instances of the lagg, in which drainage from the bog on the one side and the mineral soil on the other meet in a wet depression, that bears a more or less

Plate *12*. A series of swallow-holes in part of the raised bog complex of Komosse. H. Osvald is standing inside one of the funnels leading to the underground drainage and E. du Rietz is on the adjacent bog surface. (1958)

fenny aspect on account of the better supply of bases in the inflowing drainage water. The continuing upward growth of the bog sometimes causes the surface-water channels, the soaks, to be overtaken by later ingrowth from the sides so that, open at one place, they disappear beneath the bog surface, very much as the smaller creeks in the middle stages of a growing salt-marsh are also overgrown by the silt-retaining plant cover of the banks so that they vanish below the surface, to emerge again lower in the drainage system. Upon the surface of the big bogs near Athlone one could detect the course of such a buried watercourse by the evidence of a string of 'swallow-holes' across the bog. These are steep-sided funnels through the bog crust, 3 to 6 ft (1 or 2 m) across, and at the bottom of which one can often see open water. On one occasion I was persuaded to climb down inside one of the larger of these swallow-holes, and when I was standing precariously upon the bare peat just above water-level my raised hand was only just visible at the bog surface. One is the less likely to wander into such swallow-holes by chance because the local strong improvement in bog drainage causes their margins to be clothed by unusually tall and shrubby vegetation, including such aliens to the acid bogs as blackberry or hawthorn, as well as vigorous sweet gale, bracken, ling, purple moor grass and the royal fern.

Osvald had come on this excursion provided with a portable Hiller peat sampler, which is in essence a double-walled steel cylinder, 50 cm long and about 2.0 cm internal diameter, that can be pushed down empty to any desired depth in the peat by means of a series of steel attachment rods, and that can be simply rotated to fill it with a vertical sample of peat through the whole 50 cm space. By counter-rotation the chamber is closed again below ground, to be opened for inspection and sampling when it has been lifted to the surface. By thus examining successively deeper samples it is possible to recover the stratigraphy of peat deposits up to as much as 10 m or more, although sands, silts and clays are in general unsuitable for its use.

This Swedish peat sampler, and variants of it, became the mainstay of a very large proportion of the survey work in the first two or three decades of growing interest in peat bog stratigraphy throughout north-western Europe, and certainly in the British Isles. Here in Ireland Osvald employed it to recover for us at one or two sites the outline history of origin and development of the big and varied bogs. I give

Plate *13*. Hiller peat sampler being used to sample deposits of the former Meare Pool, near Godney, Somerset. The instrument is being withdrawn, complete with sample. (1947)

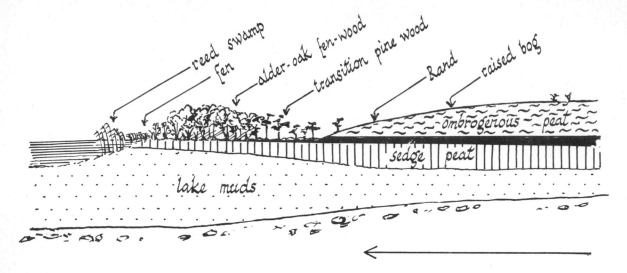

Fig. 6. Development of raised bog by invasion of open water. A floating mat of reed-swamp advances into the lake and is followed by fen which becomes tree-colonised. As the rising peat surface becomes acidic the fen-woods are replaced by pine woods on the floor of which *Sphagna* develop, and as the mosses water-log the floor, treeless raised bog takes over. The direction of development is shown by the arrow: its reality is proved by the stratigraphy. (After Steffen.)

below the condensed account of the peat sequence in the centre of the large raised bog south of Athlone.

Depth (m)

(a) 0–4.55   *Sphagnum* peat, mostly *S. papillosum*, *S. rubellum* and *S. cuspidatum* in alternating layers of low and high humification (regeneration cycle). Some *Narthecium*, *Eriophorum vaginatum*, *E. angustifolium*, *Calluna*

(b) 4.55–7.25   Fen peat, mainly *Carex* with *Cladium* and *Phragmites*: at some levels dominant *Cladium* and *Phragmites*. Muddy at base

(c) 7.25–7.30   Yellow, fine-detritus aquatic mud with *Phragmites* and chalk

(d) 7.30–9.10   Cream-coloured calcareous mud, clayey at base

(e) 9.10–   Bluish-grey clay

    This stratigraphy shews in the clearest terms that the bog originated in calcareous shallow water that became overgrown by reed-swamp and sedge-fen before passing into the ombrotrophic (rain-fed) phase in which the acidic peat of the regeneration cycle accumulated, and the typical shape of a domed bog was developed. A similar story to this can be recovered very easily from many of the raised bogs of the Central Irish Plain, where the wide exposures of the Carboniferous Limestone and glacial till affected by it must have responded to general water-logging by extensive development of wet calcareous fenland.

However in many places also, as Osvald was able to demonstrate, the sequence differs. Thus in the raised bog 3 miles (5 km) south of Edenderry, my field notes, copied at Osvald's dictation are as follows:

Depth (m)

(*a*) 0–3.25 Regeneration complex peat, mostly layers of *Sphagnum rubellum* or *S. magellanicum* peat, little humified, alternating with well-decayed dark layers of *Sphagnum–Calluna* peat

(*b*) 3.25–4.6 *Sphagnum–Calluna* peat full of *Eriophorum vaginatum*

(*c*) 4.6–6.0 Above, a *Carex* peat with *Sphagnum*, but the main mass *Carex* peat with *Equisetum* (horse-tail), *Cladium* and brown mosses (*Amblystegium*)

(*d*) 6.0–6.4 *Carex* and brown mosses

(*e*) 6.4–6.8 *Calluna–Sphagnum* peat, highly humified

(*f*) 6.8–7.8 *Carex* peat

(*g*) 7.8–8.7 Woody peat with birch (*Betula*) and alder (*Alnus*). Towards the base *Phragmites*

(*h*) 8.7– Clay

Here we find no indication of the primary colonisation of open water: rather we recognise the transgression of water-logging across mineral soil bearing wet woodland, that thus becomes sedge-fen (*f* and *d*). After a short preliminary, but impermanent move towards ombrotrophic acid bog (*e*), the wet sedge-fen eventually is converted to characteristic raised bog, possibly by way of a transitional cotton-grass community. The great majority of instances of the stratigraphy of deep raised bogs in fact fall into one or other of the two categories we have briefly indicated, although there are naturally many variations from the main trends, the most obvious being the occurrence of a layer of wood peat, indicative of fen-carr stage, at the transition from fen to acidic *Sphagnum* bog.

Tregaron, 1936–7

It is a measure of the impact made upon me by the demonstration in Ireland of the Hiller peat borer, that immediately afterwards, having failed to get an adequate copy made in the local workshops. I successfully petitioned The Royal Society to obtain a Swedish-made one. This they did, loaning the instrument more or less whole-time to me, and it was extensively used and borrowed for several years by many recruits to bog studies and to pollen analysis. The Irish excursion, in all its aspects, provided a great stimulus for one to undertake examination of raised bogs in other parts of Britain, particularly England and Wales, both in the interest of securing from them long series of samples for pollen analysis, and of examining the hydrology and surface ecology of an unfamiliar vegetational formation. At least equally alert to the potential of this opening field of research was a young graduate of Trinity College, Dublin, who was at this time acting as research assistant, courier and disciple to Professor Knud Jessen of Copenhagen, who at the request of the Royal Irish Academy had begun in 1934 and 1935 an

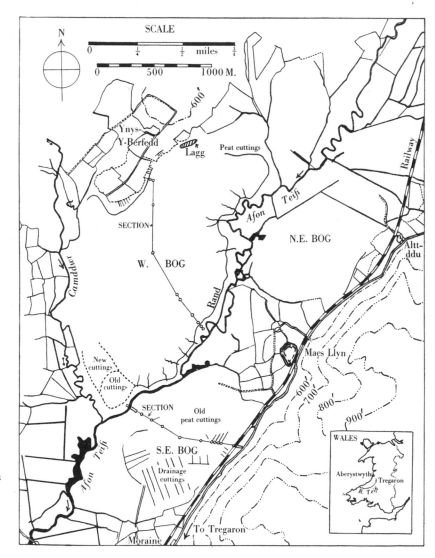

Fig. 7. Sketch-map of the Teifi valley in central Wales with three large raised bogs and the meandering river between them. Lines of section made by borings across the western and south-eastern bog are shown. See also Fig. 10.

Plate *14.* View from the adjacent hillside across the valley of the River Teifi near Tregaron, central Wales. On the near side of the river is the south-eastern raised bog, intact centrally but invaded marginally by peat diggings: at the right corner is the remnant of a fen-wood with sallow representing the lagg. Across the winding river there is the toe of the large western bog, and beyond this the far hillsides. (1936)

extensive programme of survey and research upon the Late Quaternary deposits of Ireland. This work, though interrupted by the war, was driven forward by Jessen's own dynamism as well as that of his new pupil G. F. Mitchell, who was himself also to become Professor of Geology in Trinity College, Dublin, and to achieve wide scientific distinction. We happily encountered both Jessen and Frank Mitchell during the Irish excursion, and Frank and I agreed that it would pehaps be helpful if the still unfledged recruits to bog investigations could meet somewhere and jointly try our pooled experience by working upon some peat mire in England or Wales not as yet scientifically studied. The group consisted of Frank Mitchell and myself, Dr Kathleen Blackburn of Armstrong College, Newcastle-upon-Tyne, Dr H. A. Hyde of the Botanical Department of the National Museum of Wales, and Dr A. G. North, the Welsh geologist. We chose to work upon the complex of three raised bogs that occupies the wide valley of the River Teifi in Cardiganshire (Dyfed), upstream of the terminal moraine on which, in part, stands the village of Tregaron. We accordingly spent the first half of July 1936 in a stratigraphic survey of two of the three bogs, making use of the new Royal Society peat borer and the variety of spades, trowels, sampling-, levelling- and recording-devices that these exercises call for. We taught one another the tricks of the trade, the conventions used by von Post of grading the humification of the peat, recognition of peat and sediment types and of seed and fruit remains fished out of the peat drill. Frank, I recall, taught us in the laboratory how Jessen digested the peats or muds in dilute nitric acid so as to set free their content of plant remains from the obscuring colloidal matrix and three of us, Frank, myself and Hyde, secured separate series of samples for pollen analysis.

Our reconnaissance of the complex shewed the three mires to have all the characters of typical raised bogs. The combination of drilling and inspection of marginal peat faces enabled us to reconstruct levelled sections across both the south-eastern and the western bog. They had a domed shape, sloping gently to the contact with the hillsides to east and west respectively but with steeply curved margins to the flood-terrace of the River Teifi between them. The bog centre was at least 10 m (30 ft) above the level of the river terrace, and at the hillside margin there were traces of a lagg.

In the bog centre the sequence of deposits was as shewn below:

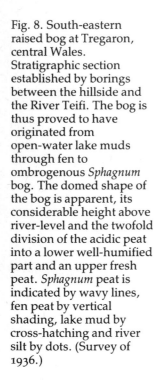

Fig. 8. South-eastern raised bog at Tregaron, central Wales. Stratigraphic section established by borings between the hillside and the River Teifi. The bog is thus proved to have originated from open-water lake muds through fen to ombrogenous *Sphagnum* bog. The domed shape of the bog is apparent, its considerable height above river-level and the twofold division of the acidic peat into a lower well-humified part and an upper fresh peat. *Sphagnum* peat is indicated by wavy lines, fen peat by vertical shading, lake mud by cross-hatching and river silt by dots. (Survey of 1936.)

Depth (m)

(*a*) 0–2.40 *Sphagnum* peat of generally low humification with some *Calluna* and *Eriophorum*

(*b*) 2.40–4.60 Highly humified dark *Sphagnum* peat with *Calluna* and *Eriophorum*

(*c*) 4.60–5.75 Transitional: as above but below with some *Phragmites*

(*d*) 5.75–6.90 Brown *Phragmites* peat with scattered wood and seeds of *Menyanthes* in lower part

(*e*) 6.90–7.25 Grey-brown mud peat with scattered wood and some *Phragmites*

(*f*) 7.25–7.60 Grey-brown *Phragmites* peat with scattered wood and seeds of *Menyanthes*

(*g*) 7.60–7.75 Brown mud peat with some *Phragmites*. Seed of *Nuphar lutea*, *Carex* fruit

(*h*) 7.75– Grey clay, sandy at top

The layers (*d*), (*e*) and (*f*) no doubt all represent an extensive reed-swamp that succeeded the open lake (of stage *g*) which had formed behind the glacial moraine in the earliest phase of post-glacial amelioration. The upper layers are characteristic *Sphagnum–Calluna* peat, divided at 2.40 m into a lower very-humified peat and an upper little-humified

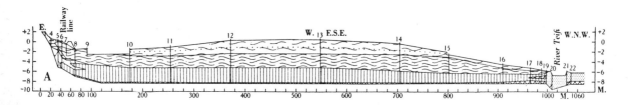

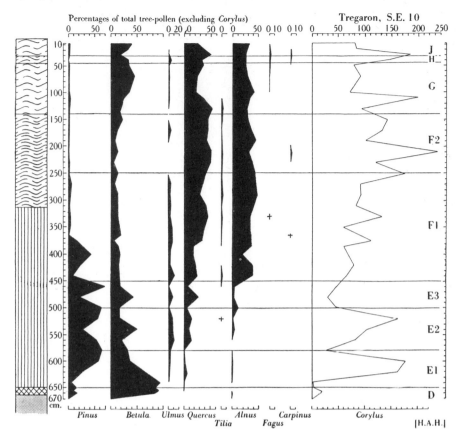

Percentages of total tree-pollen (excluding *Corylus*)　　　　Tregaron, S.E. 10

Fig. 9. Pollen diagram from the south-eastern raised bog, Tregaron, 1936. Peat stratigraphic symbols as in Fig. 38. The tree-pollen frequencies reveal the same forest succession as in all three bogs at Tregaron, and generally throughout England and Wales. Although here local pollen zones only were used, they correspond with zones IV to VIII in the national sequence (Fig. 4).

peat: this is a twofold division widely encountered through the raised bogs of north-western Europe and one which we were concerned to investigate later. The upper, pale *Sphagnum* peat we found to be interrupted throughout the centre of the two bogs by a thin layer of greasy, highly humified peat that we took to represent a phase of retardation in bog growth, possibly determined by climate. Here again was a phenomenon to which enquiry returned in later years.

In ombrotrophic mires such as these, there is much sensitivity to climatic and hydrologic change and the bog margins are especially responsive to such changes. We accordingly made many close borings and peat-face examinations in the terminal parts of our bog sections and found some evidence there of changes in bog hydrology and bog communities.

At this time, 1936, very few long and continuous pollen diagrams had been completed for Britain. The three of us who completed pollen diagrams for the Tregaron mires were happily in general agreement both as to numerical results and as to their interpretation. The deepest of the three pollen series extended from the climatic stage known in Scandinavia as the Pre-boreal: this was the time of the open-water stage of the valley. The sequence appeared to be continuous thence to the still-living surfaces of the bogs, and we tentatively equated the boundary between the pale and dark *Sphagnum–Calluna* peat with the similar European boundary between the Sub-boreal and Sub-atlantic climatic periods.

Alongside the group engaged upon bog stratigraphy there was a larger party of post-graduate and advanced pre-graduate students, whose commitment was to survey and record the ecology of the living plant communities, finding out in the process how far they resembled the Irish bog communities, and whether they seemed equally explicable in terms of the current Scandinavian concepts. We chose to work upon the raised bog on the west of the River Teifi, at once the bog least

Fig. 10. The western raised bog at Tregaron surveyed in 1937 with contours at 1 m vertical intervals: this illustrates the domed shape of the mire. The grid was the basis for our vegetational analysis and water-level observations. See Fig. 12.

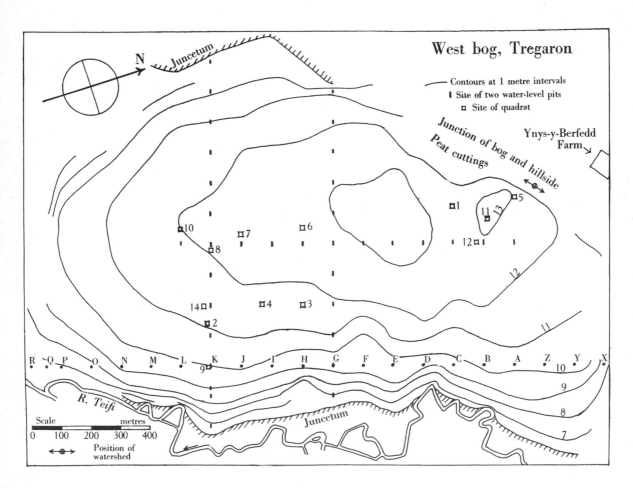

Plate *15*. The regeneration complex of the western raised bog at Tregaron, central Wales, shewing the alternation of pools and hummocks. The dark bushes are mostly *Calluna:* in several places hummock-forming *Sphagna* are invading the pools, and in the right foreground the *Sphagnum* mat carries bog-asphodel in flower.

damaged by peat cutting and burning, and the largest, about 2 km long and 1 km wide. To establish its shape beyond doubt, it was systematically levelled: a contour map at 1 m intervals exhibited its domed shape, with its highest surface 8 m above the flood-terrace of the Teifi. At the north-east the surface sloped gradually into that of the hillside and at the contact we were fortunate to find a stretch of about 200 m spared from marginal peat digging, where there still remained a modified but quite recognisable lagg. Its fenny character, attributable to the higher base status of drainage from the mineral soil, was reflected in a floristic list rich in fen species and in an extensive growth of shrubs of sallow and hairy birch, just as in fen-carr.

However our interest naturally centred in the first place upon the communities of the main area of the bog surface. Over the highest and wettest part of the bog we found a pool-and-hummock complex closely resembling that of the Irish bogs with the difference that here the chief hummock-building *Sphagna* were the brown *S. pulchrum* and the prevalently green *S. papillosum*, with the red species *S. rubellum*, *S. plumulosum* and *S. medium* (*S. magellanicum*) in a subsidiary role. Otherwise, as the diagram overleaf shews, there was great general similarity to the Irish examples already cited, and equally it was possible to make out all the stages of infilling of pools, growth of hummocks, colonisation and eventual decay.

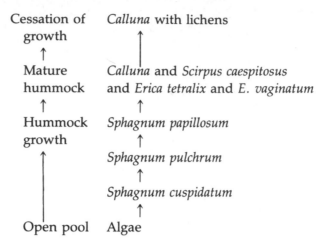

Cessation of growth	*Calluna* with lichens
↑	↑
Mature hummock	*Calluna* and *Scirpus caespitosus* and *Erica tetralix* and *E. vaginatum*
↑	↑
Hummock growth	*Sphagnum papillosum*
	↑
	Sphagnum pulchrum
	↑
	Sphagnum cuspidatum
↑	↑
Open pool	Algae

In this community, as in others, we recorded the structure in part by detailed mapping of quadrats 5-m square, charting all the main areas of plant cover. In the warm sunshine the students mostly wore plimsolls as the only protection for their feet as they paddled round on the wet

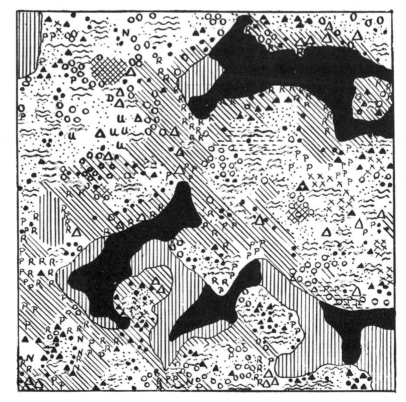

Fig. 11. A 5 m vegetational quadrat in the regeneration complex of the western raised bog at Tregaron. The open pools are black or, where occupied by submerged and floating *Sphagna*, vertically shaded. Hummock-building *Sphagna* are mainly shewn by oblique shading. Mature hummocks can be recognised between the pools: they have abundant *Scirpus, Eriophorum vaginatum, Calluna, Erica tetralix* and *Cladonia* spp., shewn respectively by wavy lines, black triangles, open and black circles, by dots and the letter u.

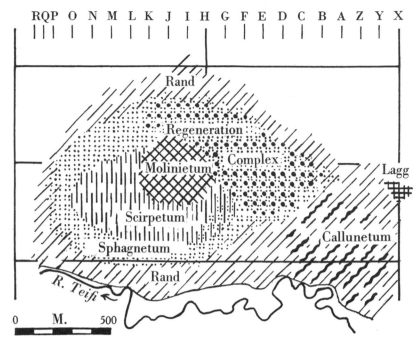

Fig. 12. The distribution of the chief plant communities of the western raised bog at Tregaron, 1937. The centre of the dome is occupied by vegetation types indicative of drying-out (*Molinietum* and *Scirpetum*) and between them and the hillside regeneration complex occupies the wettest area of the bog.

surface, and I recall the embarrassment of one party who found an adder coiled in the quadrat they had to map. It was soon borne in upon us that these bogs were much favoured by adders and I remember being startled by seeing what seemed to be several college ties, black and yellow, rippling across the bog – a female adder with her offspring! The students waved aside my protests that bare ankles, shins and calves were highly susceptible, and they scorned Wellingtons.

We were able to establish the presence of a quite typical regeneration complex, but in contrast with the bogs of Athlone and Edenderry, here large areas were occupied by communities dominated respectively by purple moor grass (*Molinia caerulea*) and deer grass (*Scirpus caespitosus*), in neither of which were pools or hummock-building *Sphagna* of more than trivial importance. It seemed that human interference might very likely have caused quite recent alteration of the bog surface, particularly by marginal peat cutting and possibly by burning. A large area dominated by *Calluna* and showing much peat erosion lay at the north end of the bog and recalled the *Stillstand* complex that Osvald had described as possibly representing the final stage of development of mature raised bogs.

We spent a good deal of time recording the character of the rand, with its typical marginal drainage streams that originate in confluent pools of *Sphagnetum* on the bog surface, cut down deeply through the curving

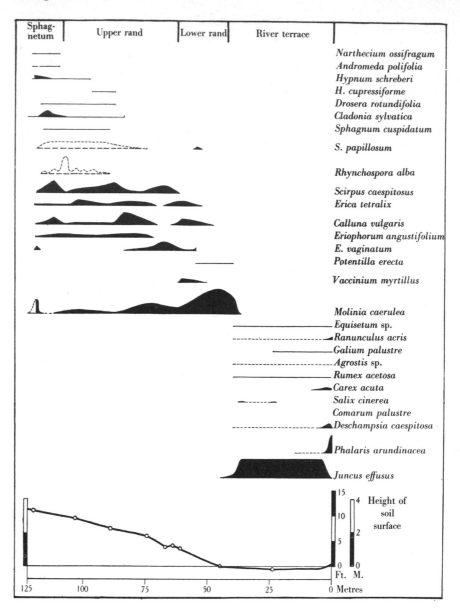

Fig. 13. A vegetational transect across the rand of the western raised bog at Tregaron, shewing the large changes in presence and coverage of the various species as one descends the bog margin and meets the river flood-terrace.

bog margin and lose themselves finally in the river terrace. We were less concerned with the river terrace, subject as it is to repeated river flooding and ecological conditions quite other than those of the bog itself. Even so the ecological work directed by Dr V. M. Conway and myself required a second season for completion, and in 1937 we made some preliminary attempt to ascertain the water-levels within the various communities of the great west bog. We were able to shew how at that (summer) season the regeneration complex and *Sphagneta* were the wettest areas of the bog surface. We also established the curious fact of an apparent discontinuity beneath the present vegetational carpet of the bog; below this the peat was almost all made of *Sphagnum imbricatum*, a very recognisable species that now is totally absent from the bog surface. Thus we finished the bog investigation in the, no doubt healthy, conviction that we left behind substantial further problems of research.

In this chapter I have attempted to shew the vigorous reaction upon my scientific curiosity of the introduction in 1935 to the phenomena of the Irish raised bogs and how at the raised bogs of the Teifi valley field investigations of considerable scope were made during 1936 and 1937 into stratigraphy, morphology and surface ecology of this category of bog. The presence and character once established, I decided that residence in the driest part of England, so far from residual living bogs, made it impracticable to pursue close ecological study of bog communities. Nevertheless, as later chapters will shew, the stratigraphy of bogs and the value of bog deposits as productive archives of geological, climatic and archaeological history were such powerful inducements that from these earliest beginnings I have myself, and through a series of very able students, exploited many such bog deposits in the interest of Quaternary research in one area or another. Indeed I was set upon this course immediately the Tregaron work was in train by first visits to the Somerset Levels in 1936 and 1937, and by an investigation in the autumn of 1938 with D. H. Valentine into the big coastal peat bog at Borth.

3

Raised bog stratigraphy: first steps

Among the great advantages of the 1935 Irish excursion and of the field trips to Tregaron in the two succeeding years was the development of our own ability to recognise, from its structure and fossil plant content, peat which had formed in raised bog, especially that from the regeneration complex. Osvald had already confirmed, during a brief visit to Cambridge, that a peat sequence preserved in a vertical monolith that M. H. Clifford and I had taken from Woodwalton Fen clearly represented the transitional stage of development to raised bog from fen, through the wooded stage of fen-carr, to acidic regeneration-complex peat. As Clifford and I set out in publications in 1938, this carried with it confirmation of evidence from other sites in the Huntingdonshire fens, at Trundle Mere and Ugg Mere, that raised bogs must formerly have been present in the East Anglian Fenland, with their characteristic fauna and flora, so alien to those of the drained and shrunken peat lands that remain there today. It has been explained in my Fenland book (Godwin, 1978) how this conclusion has proved conformable to evidence from many diverse sites in present-day Fenland. Some of this evidence is dependent chiefly upon the dominance of *Calluna* pollen and *Sphagnum* spores and leaves, in buried peat layers (e.g. at Nordelph), some based upon the relict occurrence of plants or animals typical of acid bogs (as sundew and sweet gale at Wicken Fen, and *Calluna* at Holme Fen) and even some from place-names pointing to former presence of denizens of acid bogs, such as Cranberry Farm in the South Lincolnshire fenland.

It was a particularly useful step forward that by this new information we were able to rescue considerable tracts of interesting vegetation from rather vague ecological description and recognise them as raised bogs more or less modified by drainage, peat cutting, burning and even by grazing or colonisation by sea-birds. This was notably true of the extensive peat-covered areas of the Somerset Levels, the low-lying flat land between the Quantocks, the Polden Hills and the Mendips, that have occupied my interest ever since 1936, and always with sustained reward. Until 1936 the peat-land vegetation had been described in

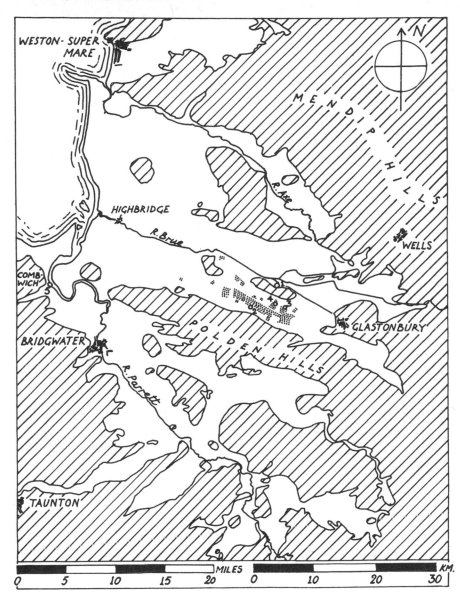

Fig. 14. Sketch-map of the Somerset Levels with the most obvious areas of uncut or partly cut relict raised bogs recognised by 1941; these are dotted with sites of early investigation numbered: (1) Meare Lake Village, (2) Shapwick Heath, (3) and (4) Meare Heath.

ecological terms only by C. E. Moss, who had referred to it in 1907 as 'heath', a very generalised term for vegetation dominated by ling or the heathers, growing upon acid soil, more or less peaty.

I was not, however, first drawn to the Levels by thoughts of the heaths, but by a notice in the *Manchester Guardian* of the date when the annual excavation would be resumed at the Iron Age lake village of Meare. Since I was hopeful that such a site might yield a pollen sequence in which this definite archaeological horizon could be precisely inserted,

I wrote to the organisers and duly had their consent to participate. They were Dr A. Bulleid and H. St George Gray, who had already been very notably associated with the excavation of the Glastonbury Lake Village site discovered when Bulleid was a medical student on vacation at his Glastonbury home. They made a formidable combination: Bulleid was a man of strongly scientific natural bent with decided geological as well as archaeological interest and St George Gray imported to their joint field work methods of exact and detailed recording such as were at that time new in British archaeology. After the resounding success of their Glastonbury Lake Village publication, they were now engaged in excavating two sites of rather similar age on the southern edge of the former Meare Pool, near to its exit over a limestone ridge at the village of Westhay. They made me very welcome and I fell into immediate rapport with Bulleid who at once saw the purpose and potential of our pollen analysis and peat investigations. Low circular mounds in the pasture appeared to indicate the position of the floors and hearths of the village huts, raised by successive layers of ash and clay, extending laterally into a black clayey soil. This soil is the occupation layer, which yielded artefacts from about 300 B.C. until well within the Roman period. The occupation layer was sealed by about 60 cm of silty lake clay and itself overlaid some 20 cm of very compressed black peat, that inspection both in the field and laboratory shewed to be a very decayed *Sphagnum–Calluna–Eriophorum* peat, embedded in which were many substantial beams of wood (some squared). Beneath this evidently ombrogenous peat was about 30 cm of compressed *Phragmites* peat that in turn overlaid amorphous lake mud. These observations at once indicated a difference from the Glastonbury Lake Village, which had been built upon a floating mass of wood peat over a deep detritus-laden lake: the raft was consolidated with much horizontal timber and brushwood and strengthened by vast numbers of vertical stakes, especially in a marginal palisading: a causeway linked the island to firmer ground. By contrast, at Meare the settlement appears not to have been floating but to have sat upon the dry margin of an acidic bog where this had encroached upon the shallow lake margin. This therefore was in no sense a 'pile-dwelling' but resembled rather the Irish crannogs, some of which remained in occupation into historic time. A similar construction has subsequently been proved for some of the prehistoric Swiss lake villages formerly also initially regarded as true pile-dwellings built over open water.

Early in my visit to the site I was given the news that one of the local turf cutters employed in the excavation had, a few months before, made the discovery of a small hoard of Roman objects in the peat he was cutting on the nearby Shapwick Heath. I begged off the services of the

Plate *16*. The peat-cutting industry on Shapwick Heath, *c.* 1948, shewing successive operations: (*a*) foreground, cubical freshly cut 'mumps' of peat, (*b*) mid-distance left, the mumps sliced into three turves along the bedding, (*c*) turves piled into small 'ruckles' for drying, (*d*) turves transferred to open stacks to finish drying and wait collection. All the cut is within the lower well-humified 'black' peat, the upper 'white' peat having already been removed.

finder, Percy Mullins, for the afternoon and with Bulleid visited the find site. It was to prove a momentous occasion for it was at once apparent that Shapwick Heath was a large raised bog, no longer actively growing, and now partly clad in well-grown woodland much of which, however, had very recently been cleared to allow peat cutting. It was in such an area that Mullins had made his discovery of a handled pewter cup wrapped in dry grass and enclosing a small earthenware beaker containing 120 silver coins that upon examination suggested a burial date about A.D. 410. Although we made some examination of the stratigraphy exposed in the peat cuttings and took a long series of pollen samples by the Hiller borer, the site was only described at all fully when I returned a year later with J. S. Turner, a Cambridge research student and close friend, who was soon to become Professor of Botany in Melbourne, Australia. We put down a levelled line of deep borings across the Heath and so constructed a stratigraphic transect of it. Southwards the line ended in the unmistakably curved margin of a rand, no doubt over-steepened by drying out towards the marginal drained pasture land separating the acid bogs from the lower slopes of the Polden Hills. The boring taken in 1936 had shewn the following broad sequence:

Depth (m)

0–0.15   Raw humus and litter of the wood floor

0.15–1.40   *Sphagnum–Calluna–Eriophorum–Molinia* peat of low humification

1.40–3.75	Dark-brown highly humified *Calluna–Eriophorum–Sphagnum* peat
3.75–3.80	Wood peat
3.80–5.25	Grey-black *Phragmites* peat with roots of *Carex*, fruits and rhizomes of *Cladium* and seeds of *Menyanthes*
5.25–	Soft blue clay

We now confirmed this pattern for the whole length of our transect. What was particularly striking was the twofold division of the *Sphagnum* peat, readily seen in the peat cuttings here and on neighbouring bogs and entirely familiar to the peat cutters. There is an upper layer of 1 m or more in thickness of rather fresh, pale-coloured peat of very low density that is useful only for litter and other purposes that require good water absorbency. Below this is a dark purplish-brown peat of high density and cheesy consistency, which is valuable as fuel and which upon drying becomes hard and horny, and not readily water-absorbent. Occasional stems of *Calluna* or tussock bases of cotton-grass make it quite evident that this is indeed a highly humified raised bog peat, though the *Sphagna* can be recognised only microscopically. There is no doubt that in general terms this twofold division corresponds with that already mentioned for the raised bogs in Ireland and at Tregaron, and, as we have said, with a similar phenomenon throughout the raised bogs of north-western Europe. The boundary between the two peat types is usually abrupt and is referred to as the *Grenzhorizont* or Boundary Horizon. Since the raised bogs owe their existence to a suitable excess of atmospheric precipitation over evaporation, it is natural that this wide-spread stratigraphic phenomenon should have been regarded as an

Fig. 15. Profile established by borings at Shapwick Heath, 1937, through the site of the Late Roman hoard. The raised bog character of the 'heath' is established by the morphology and stratigraphy. The upper and lower ombrogenous peat are sharply separated: they rest upon transitional wood peat over deep fen peat that itself rests on the flat surface of an estuarine clay.

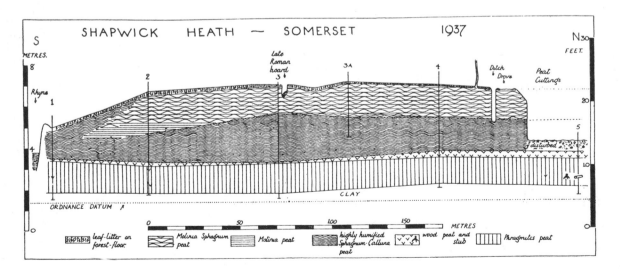

Plate *17*. Peat trench near Westhay, Somerset Levels. The upper peat here displays the strong banding resulting from compression of the pools and hummocks of the regeneration complex of a raised bog. This can also be seen in the turves set up for drying. This banded peat is in very strong contrast with the uniformly black peat, on which it rests: the discontinuity is the *Grenzhorizont*.

indicator of climatic change from the presumably warm, dry conditions of the Sub-boreal period to the cooler and moister Sub-atlantic, a shift generally thought to lie, in western Europe, at the end of the Bronze Age and opening of the Early Iron Age. An explanation in climatic terms was clearly favoured by the very widespread occurrence of the phenomenon, but a great deal of difference has been expressed about its true character, causation and age. We suspected, rightly as it turned out, that these derelict bogs of the Somerset Levels could well yield useful evidence on these issues.

The original discovery of the Late Roman hoard in 1936 was followed in 1937, and while we were there, by the find of a second coin hoard close by, suggesting a separate interment at about A.D. 388. It could be seen that this hoard had been inserted from a surface at least 60 cm above that of the 1936/7 drained bog. Two further Late Roman hoards found nearby in the next year or two also related to the same surface, which thus confirms the Boundary Horizon as substantially pre-(Late) Roman. The Meare Lake Village peat was too modified and too meagre to help our pollen-analytic project, but we were able to visit a site on Meare Heath, a mile or so away, where a Late Iron Age scabbard had been found in 1929. Percy Mullins, present when this discovery had been made, shewed us the place, but it was now so eroded and overgrown that the stratigraphic level of the find could no longer safely be recovered. Finally, however, we visited a further site on Meare Heath where, in 1936, there had been found sherds of a pottery vessel of

Plate *18*. Three pieces of pewter and a pottery vase recovered from a hoard on Shapwick Heath during 1936. The pot contained 120 Roman silver coins in which those of *Honorius* preponderated, suggesting a date of about A.D. 410 at which the objects were hidden. They were found about 60 cm below the present bog surface. Stratigraphy at the site is represented in Fig. 15. (Photograph by H. St George Gray, 1936.)

Neolithic 'B' (Peterborough) type. Their provenance was clear and it was apparent that the pot had come from the lower part of the old *Sphagnum–Calluna–Eriophorum* peat, more than 1 m below the contact with the younger unhumified peat. We had now at least established a wide archaeological bracket for the Boundary Horizon: later we were to narrow it.

Among the consequences of the work of these first two seasons in the Levels was the discovery that at every site considered, the raised bogs had originated upon the flat surface of a soft blue clay whose content of both diatoms and Foraminifera indicated an origin in brackish water. This was our first evidence, afterwards fully established (Chapter 10), that the deep valleys of the Levels had been filled by a late transgression of the sea with semi-marine deposits to a height of 1 or 2 m above present mean sea-level. Upon the extensive flats so produced there had been extensive reed-swamp before fen-woods provided firm enough footing for the new ombrotrophic raised bogs to establish themselves. As to the date when this extensive marine incursion ended, we already had the evidence that it preceded the Neolithic pot on Meare Heath. The evidence of pollen diagrams, moreover, was now consistently shewing at the bottom of the old *Sphagnum–Calluna–Eriophorum* peat a sudden and maintained fall in the relative frequency of the pollen of elm compared with that of other trees. This shift, that had already been employed as a zonal indicator (between zones VIIa and VIIb) for England and Wales as a whole, was later to be firmly associated with the earliest Neolithic forest clearances.

So now in this early stage of our tentative essay in raised bog stratigraphy it had been shewn how in quite a small area one could find

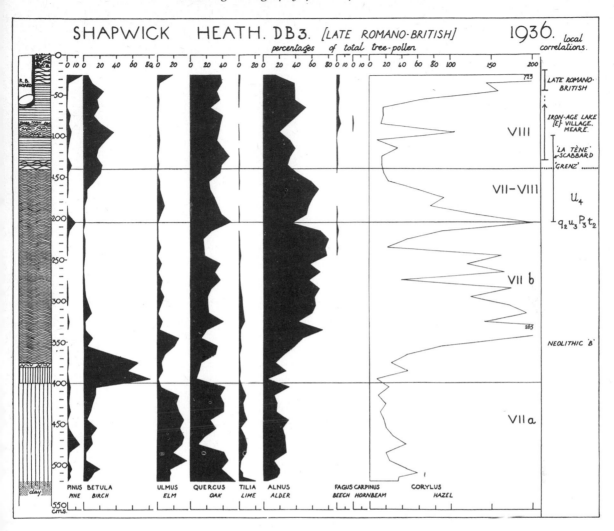

Fig. 16. Pollen diagram from the Late Roman site on Shapwick Heath, 1936. The zoning is that later introduced for England and Wales. The origin of the ombrogenous peat is close to the boundary between zones VIIa and VIIb, and the transition from the underlying sedge-fen peat is a wood-layer shown by preponderance of birch in the pollen diagram. The transition to the upper peat occurs at the opening of zone VIII where subsequent investigations shewed evidence of a major flooding episode. By the time this figure was drawn (in 1941) it had become possible to suggest the local correlations seen in the right-hand column.

the evidence of correlation between past climate, human archaeology and recent geology. No doubt we were already sensitised to the possibilities in the matter of past land- and sea-level change. At the very outset of our development of the technique of pollen analysis my wife and I had been provided with the deposits found in a series of borings made in Swansea Bay or on its margin with the great Crymlyn Bog. The *Sphagnum–Calluna* peat of the base of this raised bog was seen, beneath the sand of the coastal dunes, to lie just below Ordnance Datum (OD) upon deposits of grey estuarine clay. At various depths extending to 50 ft (15 m) below sea-level these clays were interrupted by thin compact peat beds formed in fresh-water conditions. Our pollen analyses shewed that most of this series was formed in the middle or later part of the Boreal climatic period: especially between −50 and −20 ft (−15 and −6 m), the marine transgression indicated by the sequence must have been very rapid. Thereafter the sea-level must have risen much more slowly. It seemed likely to us that we were here approaching the end of the great eustatic rise in level of all the oceans of the world, that had been induced by melting of the extensive continental ice-sheets in the climatic amelioration after the most recent Ice Age. It was of course an effect of very great scientific significance since the sea-level rise, estimated to be about 300 ft (90 m), had submerged the entire

Fig. 17. Diagrammatic representation of the submerged peat beds recovered during 1931 from Swansea Bay. Fresh-water peat formed more than 50 ft below present sea-level, and raised bog peat consistently grew at a level well within reach of present-day tides. Pollen analyses of these peats afforded an early estimate of the date of the relative land- and sea-level movements thus shewn: the oldest peat was from zones IV and V.

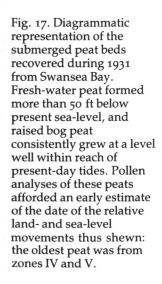
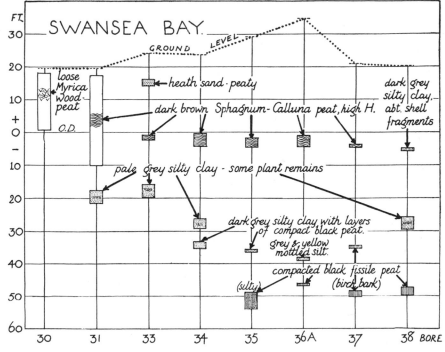

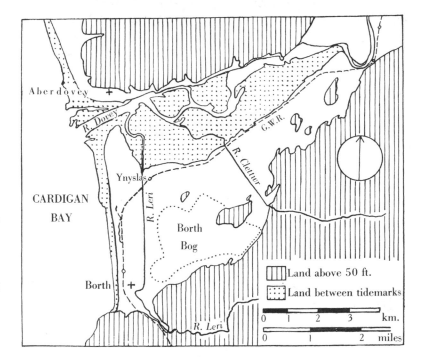

Fig. 18. Sketch map of the River Dovey estuary. The submerged forest occurs on the foreshore northwards from Borth and a ridge of shingle and sand-dune separates this from the very extensive raised bog on the landward side.

North Sea basin, that had been up till then dry land over which plants, animals and man had been able freely to move between Britain and the continental land-mass. Ireland had then also been in dry-land continuity with Great Britain, but now the British Isles were indeed islands on the continental shelf. In confirmation of the pollen-analytic age suggested by our Swansea Bay results, we could point to the fact that we had also shewn that fresh-water peat secured from the bed of the North Sea in 36 m of water was likewise of Boreal age. In terms of the estimates then available from a parallel with Scandinavian results this suggested that the last stages of the rapid ocean rise were in progress round about 6000 B.C.

Since the eustatic rise must have operated on both shores of the Bristol Channel it was evident that in the Somerset Levels equally as at Swansea, coastal flats and estuaries should have been filled by marine or estuarine deposits and that raised bogs could have formed on the ill-drained surfaces of such deposits. A similar argument indeed applies to all suitable coastal situations round these islands, and it was partly on this account that in 1938 Dr D. H. Valentine (later Professor of Botany at Durham and then Manchester University) and I undertook a stratigraphic survey of one of the most famous of all the Welsh raised bogs, the large Cors Fochno, Cardiganshire (Dyfed). This mire, commonly referred to as Borth Bog, is situated behind a coastal range of shingle bank and dune on the west, and is separated from the Dovey estuary to the north by a continuous belt of salt-marshes, the uppermost taken in as pasture. Although much modified marginally and cut through by an artificial channel of the River Leri, the centre still retains a very large area of untouched raised bog vegetation. The interest of the site was enhanced by the fact that along the whole stretch of foreshore between Borth village and the headland of Ynyslas there were exposed samples of a very well-known 'submerged forest' where both prostrate and rooted trees of alder, pine, birch and oak of large dimensions and embedded in peat, now were subject to the continual wash of the tides. Some very early attempt had been made to apply pollen analysis to this shore peat and the unfinished research already pointed to a possible link between the submerged forest on the seaward side of the coastal shingle ridge and Cors Fochno on the landward side.

Our families once established in a seaside cottage at Borth, David Valentine and I set about the bog stratigraphy, carrying I recall, the usual load of spade, peat auger and extensions, telescopic level with tripod and staff, and rucksacks with sampling gear, knives, trowels, collecting tubes, note-books, sandwiches and what have you. We crossed the River Leri by foot-bridge, only to encounter a wide artificial drain full to the brim with tidal water. We could just throw our gear across it, but for the rest it was a matter of undressing and swimming over, and dressing again in the misty rain. A memorable beginning, soon however forgotten in the interest of setting out sequences of borings across the bog, especially close in the more sensitive areas where it abutted on the Dovey estuary northward and on the drainage from the hills bounding it to the south.

Fig. 19. Pollen diagram through the foreshore peat at Ynyslas. The sequence of maxima of the tree-pollen types (oak with hazel, alder, birch, pine) was shewn to reflect the local woodland succession and the sequence of timber in the peat bed itself.

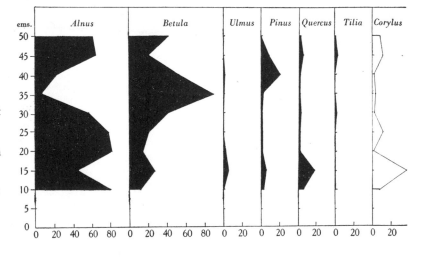

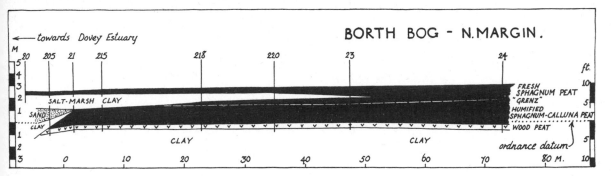

Fig. 20. Section through the northern part of Borth Bog parallel with the (canalised) River Leri in 1938. The raised bog is demonstrated to rest on the surface of an estuarine clay, and to begin with a wood-layer corresponding fully with that of the foreshore peat. Above the wood-layer the ombrogenous *Sphagnum* peat shews the usual division into a lower highly humified and an upper less humified peat with a clear boundary between. Within the upper *Sphagnum* peat a late marine transgression is registered by an incursion, from the estuary, of salt-marsh clay and associated plants.

Several conclusions of interest were established. We were able to shew that the whole bog, as well as the foreshore peat, rested upon the surface of blue silty clay at approximately 0.6 m OD. Foraminiferal analyses shewed the top of this clay to have formed in brackish water and attempts to bore through it met neither deeper peat nor hard rock. Everywhere above it was a thin *Phragmites* peat which gave place to a well-developed wood-layer that was itself directly succeeded by some 3 m of very well-humified *Sphagnum–Calluna* peat. An east–west levelled line of borings through the overlying deposits linked the bog to the submerged forest and it was quite clear that the latter was in effect only the basal part of the raised bog left exposed to seaward when the present stony beach had been driven inward across the bog margin. We had already concluded from the older pollen data that the wood-peat of the submerged forest represented the normal vegetational succession as fen-woods became more acidic and gave place to *Sphagnum* bog. The sequence of dominant trees – alder, birch, pine – is one very familiar in the base of the raised bogs of the north-west German coastal region, and in fact beside the shore of the southern Baltic the actual living fen-wood vegetation had been described. We now encountered this sequence in the pollen diagrams through the wood peat of Borth Bog, and confirmed the character of the wood peat in the sections exposed at low tide along the banks of the Leri, where quite substantial trunks of pine testified to the former presence of pine–birch *Übergangsmoor*, a transitional community that seems unrepresented in Britain today.

It appeared reasonable to interpret the situation at Borth as directly comparable with that in Somerset, i.e. as one shewing the latest stage of the great eustatic marine transgression following restoration to the oceans of the reserve of water that had been locked up in the great ice-sheets of the Glacial period until the post-glacial warmth swiftly melted them. Here again the pollen analyses indicated the subsequent fresh-water stage of fen-wood and raised bog to have begun in zone VIIa, that is to say, in the Atlantic climatic period. We were naturally

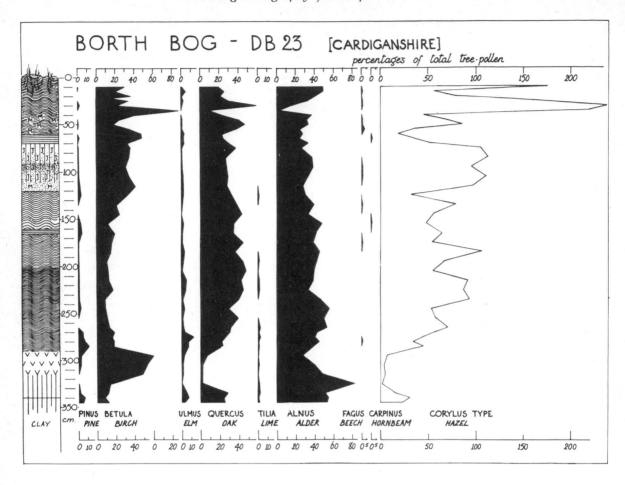

BORTH BOG – DB 23 [CARDIGANSHIRE]

concerned, having got so far, to consider what evidence there was, if any, of the subsequent much smaller marine transgression by which the base of the bog had come to be below mean sea-level and the wooded peat layer had come to be exposed within present tidal range. This evidence we found in the contact zone between the Dovey salt-marshes and the raised bog. Here borings disclosed a substantial wedge of silty tidal clay extending into the acidic peats of the bog, and very readily identified by the stout rhizomes of the sea-rush *Juncus maritimus*. The ombrotrophic peat at Borth displayed only a dubious separation into an upper unhumified *Sphagnum* peat and a lower dark well-humified one, but so far as the Boundary Horizon could be recognised, it seemed to precede the layer of tidal clay that represents the latest marine transgression. The pollen diagrams did not oppose this indication of Sub-atlantic date for it.

Whilst we were deliberately refraining from making any attempt to

◀

Fig. 21. Pollen diagram from a bore-hole in the Borth Bog profile (Fig. 20): alongside is the detailed stratigraphy with the usual symbols for peat types. Note *J (Juncus maritimus)* in the upper silty clay. The tree-pollen reflects at the base the same local forest sequence as that in the foreshore peat. Although unzoned the diagram demonstrates that it probably extends from zone VIIa (Atlantic), through VIIb (Sub-boreal) and VIII (Sub-atlantic) to the present day. The dating by this means was achieved for the relative changes of land- and sea-level: confirmation by radiocarbon dating followed.

describe the plant communities of the living bog surface, it was apparent that much of it was a low-relief type of pool-and-hummock complex and we recognised the characteristic colonies of the bright brown *Sphagnum fuscum*, a northern moss that is a much more characteristic hummock-former of continental raised bogs than of these atlantic ones. It now seems that at Cors Fochno it is near its southern limit in Britain.

The three brief accounts we have given of our first ventures into the field of bog stratigraphy, at Tregaron, Shapwick and Borth, will have made clear to the reader one of the constant qualities of this kind of research. It is the variety, not to say the multiplicity of the issues involved. No doubt because our operations are essentially an opening up of the history of events of the last few thousands of years, in effect a new way of reading the archives of this period, we have to be alert to recognise evidence bearing upon very various fields of knowledge. All raised bog stratigraphy is inherently concerned with the nature of the past vegetation types that covered it and built it up, and this is naturally associated with the former climate of the region, not only as affecting the bogs but as controlling the regional upland vegetation reflected in the bog pollen diagrams. In the identified sub-fossil plant material there is direct evidence, floristic rather than vegetational, of the past presence of species no longer present or with changed occurrence. The same may hold for animal remains. As we saw at Shapwick, and were to find increasingly, this could also hold for archaeological objects found stratified in the peat and thereby brought into relation with the environment and period in which they were in use. In all three sites we have seen a notable link with the geological history of the region, in the case of Shapwick and Borth allowing reconstruction where before it had been highly conjectural.

It is of interest to note that the discoveries at each of the three sites were such that return visits for extended research were made at all of them as experience grew, as research techniques were improved, or as chance discoveries presented favourable opportunities for fresh investigation.

4

Blanket bog

Ireland, 1935

When I had the exceptional good fortune to accompany Professor Tansley and Hugo Osvald to Ireland in 1935, I was, as will now be apparent, very greatly impressed by the vast extent of the living raised bogs and by the potentialities such mires offered alike for ecological and stratigraphic research. In point of fact the demands of our time-table took us first not to the raised bogs, but further west, into the more oceanic regions of Connemara and Mayo where ombrogenous blanket bogs dominate all flat and gently sloping parts of the landscape, which is to say, the larger part. Our engagement was to meet at Clifden, near Roundstone, Dr Robin Lloyd Praeger, most erudite and charming of naturalists, whose books *The Botanist in Ireland* (1934) and *The Way that I Went* (1937) are treasured possessions. In the year before our visit there had been brought into effect a project he had long advocated, that of instituting effective research into the history of the fauna and flora of Ireland through study of all those Post-glacial deposits in which organic remains are preserved, not least the innumerable Irish bogs. He had been made Chairman of a very powerful body, The Committee for Quaternary Research in Ireland, on which were represented the major scientific institutions of the country and particularly those most concerned with geology, zoology, botany and prehistoric archaeology. It was an organisation remarkably parallel with the Fenland Research Committee of East Anglia, though of quite separate spontaneous creation and concerned with a vastly wider territory. To launch the investigations it envisaged, the Committee had the wisdom and good fortune to engage the services of Professor Knud Jessen, who had an unparalleled record of botanical-geological research upon the inter-glacial, late-glacial and post-glacial deposits of Denmark. He was aided in his Irish researches by Hans Jonassen who acted as assistant, and by two trainee students, G. F. Mitchell and T. Maher. It was an unforeseen stroke of good fortune that, having completed one summer's research in

Plate *19*. Blanket bog in the northern Pennines near Moor House. About 2000 ft (610 m) looking west to the crest of the Pennine ridge. The most conspicuous components of the vegetation are ling, cotton-grass (especially *Eriophorum vaginatum*) and deer grass (*Scirpus caespitosus*).

Ireland, the team should chance to be in Roundstone at the time of our visit, where indeed the Chairman of their Committee was meeting them. This encounter began for me two of the warmest and most productive friendships of my life, for both Knud Jessen and Frank Mitchell shared my own enduring enthusiasm for Quaternary research and both attained the highest international recognition therein. Their quickness of insight was matched by extraordinary capacity for sustained careful work and by a sense of humour and good fellowship that ensured the success of the many arduous field tasks they engaged upon, and with which they involved many and various helpers.

Our visit to the far west took us into territory where great stretches of landscape were covered by acidic ombrogenous blanket bog, extending as a mantle of saturated black peat up to 6 ft (2 m) or more in thickness, over everything save steep hillsides. These mires lacked the decisive morphology of the raised bogs and where they turned to slope up the hillside they exhibited no sign of a wet lagg or sloping rand. Although the plant species of the bog surface were the same as those of the raised bog, there was little sign of a regular regeneration cycle from pool to hummock and back again, though here and there a massive solitary hummock would protrude above the general surface. The pools were all small and shallow and the bog mosses, the *Sphagna*, clearly played a far less significant role here than on the raised bogs. A general but variable dominance of the bog communities was shared by five species: cotton-grass (*Eriophorum vaginatum*), purple moor grass (*Molinia caerulea*), deer

Plate 20. Drainage pool on flat surface of western Irish blanket bog, with floating *Sphagna,* bog-bean (*Menyanthes*) and beaked sedge (*Rhynchospora alba*). Left to right, H. Gams, K. Jessen and G. Negri. (1949)

grass (*Scirpus caespitosus*), beaked sedge (*Rhynchospora alba*) and black bog rush (*Schoenus nigricans*). All of these are monocotyledons and help to give the aspect of an untidy and rather drab sward, but one whose uniformity is relieved by abundant bushes of ling and cross-leaved heath (*Erica tetralix*), and often by sheets of the yellow bog-asphodel (*Narthecium ossifragum*) in the shallow pools, along with red-rattle (*Pedicularis palustris*), bog-bean (*Menyanthes trifoliata*) and, at the margins the sparkling rosettes of sundews (*Drosera* spp.), or the less frequent butterwort (*Pinguicula vulgaris*). The *Sphagna,* though playing a lesser role than on the raised bogs, are the same species, except that those of the section *subsecundum* here frequently occupy the open water along with *S. cuspidatum.* The other Bryophyta are again those of the raised bogs, but the larger hummocks are very commonly crowned by a moss, the shaggy *Rhacomitrium lanuginosum,* whose white curling leaf apices give its tussocks a typical woolly appearance that sorts well with the tolerance of extreme conditions displayed by this species.

In these western areas there is a profusion of shallow lakelets over the bog surface, possibly residual to some extent from the uneven surface of the glacial till on which the peat bog has grown, and Osvald was able to shew us how sometimes such lakelets are linked by small streams or slower 'soaks' that cross the mire surface, and that may be transformed into a string of swallow-holes as in the raised bogs. In climates as oceanic as that of western Ireland it is inevitable that there should be a great deal of free water standing upon the mire surface and there is considerable minor water movement giving locally improved drainage, accompanied by greater importance and more robust growth of plants such as *Molinia* and *Calluna*. What is perhaps surprising is that upon slopes as great as 10, 15 or even 20 degrees, the climate should be so favourable to peat-forming plants that their uniform carpet is kept intact by continuous growth. It is ironic that this excess of water reduces the blanket bog countryside to a desert condition just as effectively as does the lack of water in the world's driest regions. The general poverty of the vegetation was evident in the life standard of the peasant population: it was not uncommon at this time to see a single pig being driven along the bog road by a barefoot man carrying his boots to save the soles from wear.

Despite the generally meagre quality of the acidic saturated peat, a few features in the vegetation require us to make some qualification. The black bog rush, *Schoenus nigricans*, here often present in high frequency, is a plant that elsewhere in the British Isles is closely associated with calcareous fens, and this has led to the hypothesis that wind-borne spray from these exposed Atlantic shores may well carry mineral salt in fair amount to the bog surfaces. This might also account for the presence in the bog pools of *Sphagnum subsecundum*, more characteristic of eutrophic habitats than most *Sphagna*, and could well have to do with the presence of mildly eutrophic species, such as *Cladium mariscus* and *Phragmites communis*, where streamlets drain into larger bodies of base-deficient acid water.

An alternative and often plausible way of explaining the presence of such more demanding species is by invoking the presence of water drained from mineral soil nearby. Lennart von Post had indeed been so greatly impressed by this possibility that he had created a whole major category in his own classification of peat bogs, the 'soligenous bogs', and despite the handicap of never having visited Ireland, had classified a large part of the Irish peat bogs as soligenous. Now, met on the deep quaking blanket bog near Roundstone, our party stood in a group in the fine Irish rain, listening to Jessen and Osvald thrash out the pros and cons of such a conception. The rain fell unremittingly, our Wellingtons sank slowly to calf depth in the soft bog and after half an hour or so the international conference was adjourned, the problem unsolved. As in most such cases, it seems to me retrospectively that the matter was one of semantics and usage: to some extent what one calls 'soligenous', the other calls 'topogenous', and so on. In classification there is never a single 'correct' answer: each different classification is suited to its own assumptions and circumstances and requires employment only in relation to a stated context or aim. Certainly on the one hand no ecologist was going to deny that there are recognisable soligenous influences in peat mires, and equally no one seeing an Irish hillside covered with blanket bog except for a few small crags near its crest, is going to suggest that water drained from this limited mineral exposure is the effective means of keeping the whole hillside wet enough to carry peat, which therefore is 'soligenous'. The presumption of ombrogeneity is overwhelming. What was at the time outstandingly striking to the silent junior observers of the dispute was the intense earnestness, wide experience and tolerance that both disputants displayed.

In contrast with that of the raised bogs, the peat constituting the main body of the blanket mire is featureless and uninformative stratigraphically. It is black or dark-brown, highly humified, shewing few recognisable plant remains and what there are seem mostly to be fine fibrous monocotyledonous rootlets: *Sphagna* are generally sparse or apparently absent. None the less, throughout the vicinity of Roundstone we were able to observe that this black featureless peat everywhere rests upon a layer of wood peat that commonly carries the crowns of large pine trees grown *in situ*. In its lowest layers, resting upon the mineral soil, the wood peat is mostly of birch. One is consequently led to the conclusion that originally the landscape was forest-clad, and with a climatic change

Plate *21*. Section through blanket bog in North Mayo, exposed by erosion and disclosing a large pine rooted in podsolised sand and gravel below a substantial thickness of peat, *c.* 2 m. In foreground, H. Osvald. (Aug. 1935)

towards oceanicity it became increasingly water-logged and peat-clad. It was natural to suppose that this climatic shift might well correspond with that recorded in the raised bogs at the main Boundary Horizon, and that this vegetational change had therefore happened at the Sub-boreal/Sub-atlantic climatic transition in the later part of the European Bronze Age. This indeed was Jessen's provisional conjecture and in his continuing work on Irish Quaternary history he devoted considerable attention to field evidence of artefacts of this period recovered from the blanket bogs, as subsequently did Frank Mitchell.

Of particular interest at Roundstone was the exposure of blanket bog, sectioned by marine erosion, that Jessen was able to shew us. About 1 m of *Eriophorum–Molinia* blanket peat at this site sealed in a wood-layer with pine stubs in the position of growth and below this were birch remains. As we were able to confirm by boring from this level, the wood peat overlaid *Phragmites* peat and then fresh-water lake mud down to a depth of 5 m below low-tide level. The rocks and eroded peat covered with the tidal wrack, *Fucus spiralis*, brought home visually the conclusion that since the pine forest had flourished and since blanket bog had formed to at least part of its present depth, there must have been a rise of sea-level relative to that of the coast. Here again both Jessen and Mitchell were assiduously to seek field evidence that might elucidate the timing or magnitude of this geological event. It ought of course to be said, before being too hastily carried away by the prospect of a gratifying solution to the problem of the age of the blanket bog, that this wording itself makes a misleading simplification of the issue, since one must leave open the possibility that blanket bog may well have started to form at different times in different localities and subject to locally varied conditions of slope, shelter, altitude and so forth.

The unhesitating support we gave to the view that the presence of blanket bog was due essentially to the climate, indicated our acceptance of the conclusion, afterwards explicitly stated by Tansley in his book, that for this climatic regime blanket bog is the natural vegetational 'climax', that is to say, the terminal or self-perpetuating stage upon which converge all the autogenic (self-propelled) vegetational seres, whether from dry open soils or from open water. We had, however, no means of judging how far the blanket bog might be able to persist and regenerate: the three millennia we gave as a rough estimate of age was not long in terms of the processes involved; it was, however, long enough for further shifts of climate to have modified the bog development, and most significantly we were now living in times when human interference with the bogs, as by desultory grazing or by nibbling of turf cutters at the edges, might be obscuring or frustrating the play of natural

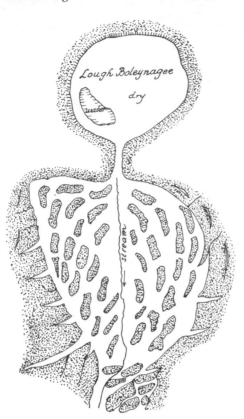

Fig. 22. Bog-burst in the blanket bog in County Mayo, Ireland, schematically drawn by H. Osvald in 1935: Plate 22 is a photograph taken at the same time.

ecological processes. Osvald had described for the raised bogs of Sweden what appeared to be terminal stages in the form of *Stillstand* and erosion complexes, in which the vegetational cover shewed breakdown to bare peat and recovery to varying degree, but we saw no convincing evidence of this in the blanket mires. On the other hand we were able to inspect, at the site of a former 'bog-burst', a type of retrogression of which there have been far more instances in Ireland than anywhere else in the world. This example was at Glencullin, near Bangor, County Mayo. The reconstructed events of the disaster appear as follows. A gentle hillside covered with blanket bog sloped upwards at about 5 degrees to a flat hilltop where peat growth had held up a small lake, Lough Boleynagee. About 1930 excessive rain had filled and overfilled the lake and the water saturated the peat of the slope below, especially along the line of a shallow valley in the underlying mineral soil. It appeared as though the very humified black bog peat, now extremely fluid, ruptured the surface crust of living vegetation and less humified upper peat, the break occurring some way down the valley at a slight increase in the angle of slope of the bog. Through this rupture the black 'porridge' of the lower peat poured out, filling the shallow valley in places to a depth of 5 m, extending downhill along the line of the most fertile soil where the original valley had deposited colluvial material. The lakelet was drained and the surface crust of the bog, no longer supported by hydrostatic pressure from below, collapsed into the path of the black liquid mud. When we visited the site in 1935 a considerable part of the hillside had been denuded of bog, but stranded upon the mineral soil were numbers of large blocks of the blanket peat, vegetation still intact. The stranded blocks were crescentic in plan and several metres long (10 to 20 m long and 2 to 3 m broad), and at the contact with the still-remaining bog there were numerous great curving splits round the open margin. Half-detached blocks shewed a strong tilt of the bog surface downwards to the side of the fracture and it was apparent that each individual raft had undergone what is known to geophysicists as 'rotational slip'. The rupture once begun, it appeared that the unsupported margins of the bog had suffered this type of shearing which extended up the slope and sideways into the bog until Lough Boleynagee had been drained and enough of the lower wet peat removed to stabilise once more the bog surface. We were remarkably fortunate to have seen a site already well documented, but still presenting so much visual evidence of the burst. Mitchell has described other sites of bog-bursts, some recent and others ancient and overgrown but still recognisable. It

Plate 22. Bog-burst in the blanket bog near Glencullin, County Mayo, photographed looking uphill. Note the great crescentic crevasses and the huge blocks of the upper peat that have fallen as the lower liquid peat flowed from beneath them. (Aug. 1935)

may be by processes of bog-burst and regeneration that a climax cover of blanket bog can be maintained.

As one moves eastwards across the British Isles one finds that blanket bogs are progressively restricted to the mountains, where elevation induces the higher precipitation and the cooler and moister atmosphere than is normal for the region. Thus in the Wicklow Mountains, south of Dublin, blanket bog is generally present above 1800 ft (550 m). When the bog was intact it carried vegetation quite similar to that of the low-altitude blanket bogs of the far west, on the flatter surfaces tending to develop a pool-and-hummock morphology similar to that of the raised bogs. Where the ground sloped more steeply, between 5 and 8 degrees, the pools and hummocks became fewer and smaller and *Sphagnum* less abundant. At the higher levels the surface was dominated by the drab sward of deer grass (*Scirpus caespitosus*) and grey *Rhacomitrium* moss. We were now in the region of the extensive peat erosion that makes so dramatic a feature of the upland mountain bogs throughout Britain. Osvald sees this erosion as originally due to wind action undercutting the up-slope margin of an open pool, proceeding the faster when freeze and thaw fracture the exposed peat, each small cliff working back until it reaches the next pool above it. Hollows thus united carry water downhill and these intermittent streamlets add their own erosive powers to those of the wind, rain and frost. As the streamlets converge and cut deeper they expose the mineral soil conspicuously along their beds and remove so much of the peat cover that in extreme cases it is left only in gaunt isolated hags,* with concave flanks of crumbling black or dark-brown peat and a topping of such plants as appreciate the improved drainage there, particularly the deer grass (*Scirpus caespitosus*), ling (*Calluna vulgaris*) and *Rhacomitrium*. Our small party was in no position to consider the basic question of whether all mountain blanket bogs suffer this kind of extensive gully erosion as a natural end-stage or whether blanket bog regenerates after it. We were content to note that the peat profiles exposed by bog erosion were essentially like those in the western bogs: about 1.5 m thick and with somewhat paler *Sphagnum–Calluna–Eriophorum* peat above, overlying a black structureless peat containing here and there remains of pine or birch. We saw pine wood at 1700 and 1900 ft (520 to 580 m) and understood this to be close to the upper limit at which it is found, though birch occurs still higher. Here, as in Connemara, one is apt to see this tree growth as evidence of a climate formerly less oceanic.

* There is an odd confusion of popular usage: in some regions the term 'hag' is applied in the sense here described, but elsewhere the 'hags' are the erosion channels themselves!

Like so many of the phenomena and hypotheses to which we had been exposed during the Irish excursion, gully erosion was to remain for many years afterwards a subject for speculation and extended observation.

England, 1935

Osvald having returned to Sweden, Professor Tansley and I, together with H. Hamshaw Thomas of Cambridge, in August of the same year visited the well-known upland moors of Cornwall and Devon to determine in the light of our Irish experiences into what category of mire Bodmin Moor and Dartmoor could best be placed. It was essentially a reconnaissance, brief but helpful. The gently rounded granitic hills proved indeed to be covered with blanket bog recognisably similar to the Irish examples, but generally more modified by human interference. On Bodmin Moor especially one could see that the vegetation had been altered by tin-streaming and, on a larger scale, by kaolin extraction, both having considerable effect on drainage. There was also much grazing. The centre of interest on Bodmin Moor was turned from the blanket bog vegetation to the remarkable stratigraphic exposures in the china-clay workings at Hawks Tor, where, instructed by contacts with Jessen and Mitchell in Ireland, it seemed to me that we had noteworthy evidence of conditions prevailing in the south-west through the climatic changes at the end of the last glaciation. A layer of large stone blocks and gravel was traceable down the hillside and out into the organic deposits of a channel in the deeply kaolinised surface of the granite, and there was no evident mechanism by which this unsorted stony layer could have moved, save by alternate freezing and thawing, the 'solifluction' process well known in southern Scandinavia to have occurred during the short spell of returned cold after the Late-glacial amelioration known as the Allerød period. We took time to examine, record and sample this important section to which we returned in the following years, and from which very useful conclusions were then achieved. In 1935, however, we turned back to our main purposes by now visiting Dartmoor and in particular the southern half, less affected by deep erosion than the northern part with the famous eerie Cranmere Pool at its heart. Cater's Beam is the gentle summit about 3 miles (5 km) south-south-east of Princetown, and just above 1500 ft (450 m) elevation. It is covered with blanket bog which forms the collecting areas for the rivers Swincombe and Strane, flowing respectively north-east and north-west, and on its southern flank is a gentle concavity in which numerous confluent tributaries drain the blanket bog and constitute the headwaters of the south-westward flowing River Plym. The general plateau bog at Cater's Beam, with slopes generally less than 3 degrees, was blanket bog with *Molinia*, *Eriophorum* and *Scirpus* roughly co-dominant but with very abundant *Calluna* and frequent *Erica tetralix*. The peat was generally 2 to 4.5 ft (0.3 to 1.4 m) thick, unhumified and with recognisable *Sphagna* only in the deeper parts. The lower and steeper part of the river-head system shewed much bare peat and pools with cotton-grass between residual hummocks of *Molinia*, *Erica* and *Scirpus*, but around this a broad band of gentler slope carried deeper peat, where among the tall tussocks of *Molinia*, *Erica tetralix* and *Scirpus* there were local depressions with abundant *Narthecium*, *Rhynchospora* and *Sphagnum cuspidatum*, and in some parts tussocks of *S. rubellum*, *S. papillosum* with *Leucobryum*, *Drosera* and *Cladonia*. Neither here nor on the plateau could one recognise a regular regeneration cycle, nor was such a cycle represented in any of the exposed peat faces. There was very little doubt in our minds of the affinity of this mire vegetation as a whole with the blanket bogs we had so recently seen in Ireland. We deliberately avoided any attempt to survey the extensive and extremely wet bog that filled the valley at Fox Tor Mire, only half a mile (0.8 km) to the north of Cater's Beam: certainly it must have been strongly topogenous and we were ill equipped for such an aquatic venture.

In the following summer, as a convenient extension of field work at the Hawks Tor china-clay pit, my wife and I visited the more extensive,

Plate 23. Blanket bog of northern Dartmoor, shewing peat diggings through the humified peat to the weathered granitic surface. The three botanists are H. Gilbert-Carter, E. F. Warburg and Margaret E. Godwin. (1936)

Plate 24. Blanket bog, north Dartmoor. This shews erosion by wind, rain and frost on the former peat diggings and suggests some possibility of natural regeneration of blanket peat over the granite, primarily through cotton-grass. (1936)

higher half of Dartmoor, and secured samples for pollen analysis near the top of Okement Hill; the peat, some 1.70 m deep, was very highly humified and quite black. The vegetation was that of blanket bog subjected to heavy drainage from extensive peat cutting, with prominent *Molinia, Scirpus, Calluna, Erica tetralix* and *Eriophorum angustifolium* accompanied by the prostrate heath rush *Juncus squarrosus* and *Festuca ovina* (sheep's fescue), both perhaps a response to grazing. *Sphagna* were almost absent.

One of the questions prompted by the sight of extensive removal of blanket peat down to the mineral soil, whether by erosion or by widespread peat cutting, is whether the bog will regenerate under conditions now prevailing, and if so, by what vegetational stages. Easily enough posed, the question is difficult to answer: one may cover big areas of heavily eroding blanket bog and find no convincing examples of restoration. Though *Eriophorum angustifolium* often flourishes on spreads of crumbly eroded peat on the floors of erosion gullies, succession seems to get little further. Here, however, on the flat bogs of northern Dartmoor, within a large area of peat removal to the mineral floor there appeared to have grown a fresh cover of peat to about half the original depth, at which stage it had apparently itself been cut out and a third thinner layer had occupied the bare ground. Thus there were shelves of flat bog with straight boundaries attributable only to peat cutting. This observation impressed me greatly and I have always regretted being unable to return to the site for proper recording and observation of it, not feasible at the time.

Now, by the end of 1936, thanks to Hugo Osvald and our Irish hosts,

Plate 25. Peat-cutting tools on blanket bog of Dartmoor. Second from the left is the wooden becket shod with iron that made the sloping cuts shown in the peat face. The other tools are a peat spade and knife: that on the left is probably a mattock for clearing the tough top crust. (1936)

we had gained some general familiarity with blanket bog and had learned enough to guess how much more difficult would be the study of blanket bog ecology or stratigraphy than the corresponding research into raised bogs, especially as in Britain at least, the blanket bogs are high in the mountains, in particularly wet climates and in localities distant from our Cambridge base. Likewise the blanket peat was relatively thin, apparently devoid of internal differentiation and less likely to have preserved artefacts illuminating the occupation of prehistoric men. At all events, for the next few years our investigations were centred upon the more accessible raised bogs, although from time to time more or less sustained attacks were made upon particular areas of blanket bog as it suited the convenience and interest of students and friends.

5

Plants of the bogs: *Sphagnum*

I have never been in doubt that the task of the ecologist in understanding vegetation is much facilitated by a close knowledge of what we can call 'the intimate biology' of the individual species that compose it. All the more must this be true as he seeks to discover the causes by which one vegetation type develops into another, or when he seeks to reconstruct past vegetational history from the sub-fossil plant remains that constitute the archives of the peat bogs. In a book of this kind one cannot make a long comment upon the 'autecology' of the bog species, but none the less I have felt it must be helpful to offer what could be called a simplified portrait of a few of the most striking and characteristic of the bog plants; they are after all the *dramatis personae* of the historic pageant presented by the evolution, maturation and decay of the peat bogs.

So preponderant is the role of *Sphagnum* moss in building the ombrotrophic bogs, and so omnipresent are its many species, living on the bog surface or sub-fossil in its strata, that one cannot do other than provide for the reader at least an outline sketch of its nature and properties.

Although individual shoots are small, as in all the mosses, they often aggregate into living colonies that gain special properties from the association, as well as thus becoming more conspicuous, as when they form the large crimson hummocks of the raised bog and confer the title of 'red bogs' on the mires they dominate. The genus is a large one and the many species typical of British mires exploit a wide range of habitats therein, based mainly upon their individual adjustment to conditions primarily of the wetness of the environment (the relation mostly to water-level), to the base status and acidity of the ground habitat and to the local degree of shading. The submerged or half-submerged diffuse green mats of *Sphagnum cuspidatum* occupy the raised bog pools, whilst the species of more eutrophic waters are those rusty-ochreous members of *S. subsecundum* group. The uppermost hummock builders tend to be densely crowded, especially so in the 'singed' brown *S. fuscum*, and in *S.*

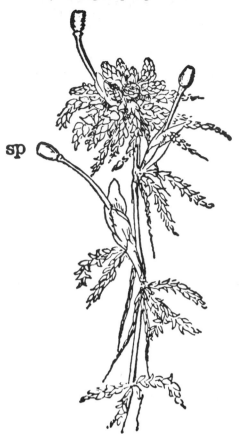

sp

Fig. 23. A species of *Sphagnum* moss shewing the typical spreading terminal branches and the descending branches that form a 'wick' beside the main shoot. There are also three of the leafless stems that bear the spore capsules.

imbricatum, its name betraying its compact leaf and shoot arrangement. The main hummock builders, *S. rubellum, papillosum, plumulosum* and *magellanicum* are all naturally aggregated into big clumps, and it is these that display the most brilliant colouration, *S. magellanicum* in particular often making a marvellous purple-reddish metallic display in its expanded leafy shoots. Relatively few species of *Sphagnum* occupy the bare peat surfaces, though these include both the mat-forming *S. compactum* and the sparse, slender *S. tenellum*. Very few of the genus tolerate shade, though here one may cite *S. squarrosum* with its leaves more or less sharply reflexed from the leafy shoot. This is also a species more tolerant of eutrophic conditions, but in respect both of shading and of high mineral concentration, the limits of the whole genus are similar and very quickly reached.

With such pronounced limitations, the one strongly operating in the competition of taller flowering plants and the other vastly reducing the extent of suitable territory, it is pertinent to point to the major competitive *advantage* held by these low-growing mosses. It rests, in the very felicitous phrase of R. H. Yapp, in the fact that by holding up water so as

Plate 26. Photograph from above of a *Sphagnum* moss carpet, with a flowering plant of common sundew (*Drosera rotundifolia*) with its crozier-shaped inflorescence and tentacle-clothed leaves. (Photograph by W. H. Palmer.)

to water-log the ground, they establish (in relation to all plants rooted beneath them) a 'priority for oxygen'. They thus can determine a vegetation type which is so dwarf and surface-rooting that it can hardly at all exercise that 'priority for light' to which the *Sphagna* themselves are so susceptible.

This competitive advantage of the *Sphagna* derives of course from their outstanding powers of water absorption and retention, powers themselves due to the habit and morphology of the plant. The shoots and leaves are of such small dimensions and aggregate so closely that a capillary film of water extends over and betwixt them, continuous with that similarly held in the spongy peat surface formed of similar but dead material. The capillary ascent of water is facilitated by the fact that a proportion of the lateral leafy shoots grow vertically downwards, clothing the primary axis of the aerial stem with an open 'wick' down to its base. Even more strikingly adaptive is the microscopic anatomy of the moss leaves themselves, all of which exhibit a remarkable configuration, readily seen under the microscope but detectable also by a good lens (Fig. 24). The mature leaf is one cell thick, presenting a netted appearance in which each mesh is a large 'hyaline' cell, empty of contents, communicating with either or both surfaces by wide round open pores and made conspicuous by slender spiral or annular bands of thickening, running round the inside wall and emphasising the inflated sausage shape of these water-storage cells. The walls of the network consist of chlorophyllous cells, living and green with the contained oval chloro-

Fig. 24. *A*, a leaf of *Sphagnum* moss seen through the microscope. The green chlorophyllous cells (c), have been hatched: they enclose the empty water-storage cells whose walls are pierced by large pores (p) and are supported by spiral wall-thickenings (s). *B*, sections of the leaves of two different species of *Sphagnum:* in one the green cells directly abut on the upper leaf surface, in the other they are buried centrally by the water-storage cells.

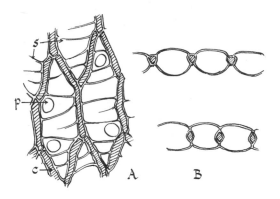

plasts, the seat of the photosynthetic activity on which autonomous growth depends. These chlorophyllous cells in some *Sphaga* (such as *S. imbricatum* and *S. papillosum*) sit between the hyaline cells but close enough to the upper leaf surface to expose themselves there, but in other species (as for instance *S. magellanicum* and *S. compactum*) the green cells are deeply buried in the centre of the leaf by the surrounding hyaline cells, out of direct contact with the atmosphere. There is no doubt of the adaptive character of the hyaline cells as organs of water-storage, presumably seeing the moss through temporary phases of drought. Equally, it is evident that the internal coils of thickening prevent collapse of the hyaline cells under hydrostatic tension and add a component of outwards wall-pressure to the water-absorbing power of the drying cells.

It is no surprise that a wet tussock of living *Sphagnum* should hold water many times the weight of its own dry tissues. Equally it is unsurprising that its water absorption should confer considerable economic value on the moss, either living or as a principal component of peat. *Sphagnum* moss has long been employed as stable litter, a role where its absorption of 'stock nitrogen' and potassium facilitates transfer of these valuable elements to cultivated ground. More recently there has grown up an immense industry supplying macerated *Sphagnum* peat to the horticultural trade, which is not slow to point out its advantages in improving both water-retention and mineral conservation in dry soils and rain-deficient regions, as well as providing an excellent basis for indoor bulb-growing, and a generally weed-free matrix in which seedlings can be raised in commercial quantity. There still also remains a strong memory of the collection, during the 1914–18 World War, of great quantities of fresh *Sphagnum* moss for use in emergency field-dressings, where they provided absorption of blood and pus together with a degree of antisepsis. It is more than likely that this was a usage of great antiquity; thus it was conjectured that a thick mass of plant material,

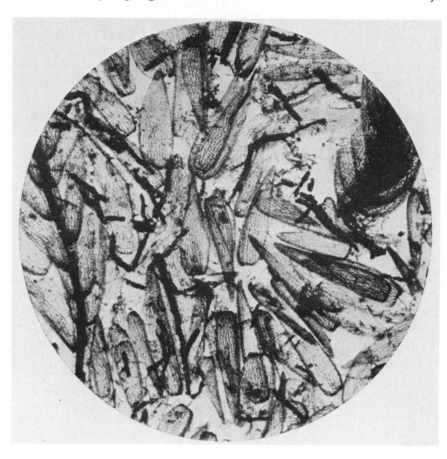

Plate 27. *Sphagnum* peat: a very slightly humified sample seen after maceration under the low-power microscope. Characteristic leaf and shoot structure are fully recognisable. (Photograph by H. Osvald, 1937.)

largely of *Sphagnum palustre,* found on the chest of a Bronze Age skeleton recovered from a grave in Ashgrove, Fifeshire, had been placed there to staunch a wound that had, however, proved fatal.

The *Sphagna* are flowerless plants that reproduce mostly by the spread of vegetative shoots and more rarely by the production of spores in spherical capsules borne above the green shoots. The spores are recognisable at the generic level and can be useful in a pollen diagram, as shewing for instance the horizon of the onset of soil acidification or ombrogenic mire. At the same time spore production is exceedingly variable and many species are seldom found 'fruiting' in this country: one of those most regularly seen is *S. plumulosum.* Thus the peaks and troughs of a *Sphagnum*-spore curve through a bog profile can be almost impossible to interpret.

Every acidic mire that is not polluted or desiccated bears a great wealth of flowerless plants in its vegetational cover, particularly of mosses, liverworts and lichens. With these, *Sphagnum* apart, I have deliberately not concerned myself, nor do they figure in these short plant portraits. This has followed from the fact that they play an apparently minor role in the ecological processes of bog-building, are preserved to a minor extent only in sub-fossil state and make a somewhat less evident and dramatic pattern to all but the *cognoscenti*.

Even within the flowering plants, no claim at all is made for comprehensiveness: the aim is by no means to present a short biological bog flora. What is intended is to catch the reader's mind by descriptions of a selection of a few of the species about which one can say enough to bring to life their role and indeed their personality in the constant progress of bog vegetation change. Just a few eligible entrants, such as the sundews (*Drosera* spp.) and cranberry (*Vaccinium oxycoccos*), I have already mentioned in a similar manner in my book about the Fenland (1978), and these are not reconsidered.

6

Plants of the bogs: sedges and such

When the monocotyledonous plants arose by evolution from the dicoty-ledons early in the history of the flowering plants it seems likely that the change was in large part associated with adaptation to an aquatic habit of life. Certainly the great monocotyledonous family of the sedges, the Cyperaceae, is particularly rich in genera and species of aquatic, marsh and bog plants. These plants seem not only to have lost the wide leaf-blades of dicotyledonous plants, now exhibiting narrow parallel-veined leaves that might derive from flattened petioles, but time and again they shew great reduction of even the narrow monocot leaf to a mere scale. In such instances the photosynthetic functions have been substantially taken over by the green flowering stems that play such a conspicuous role in plants such as the bulrush (*Scirpus lacustris*), deer grass (*Scirpus caespitosus*) and spike rushes (*Eleocharis* spp.). It is not surprising, therefore, that the Cyperaceae, with their stiff green flower-ing stems and narrow tapering leaves should seem to supply, in large degree, the 'pile' to the carpet of vegetation in the ombrogenous mires, extending alike across flat swards in local depressions and flourishing on the hummocks, alongside certain grass species such as *Molinia caerulea*, the purple moor grass, that in such conditions may forsake its usual hummock-forming habit in favour of a prostrate rhizomatous one.

White beaked sedge   *Rhynchospora alba*

In high summer our raised bogs and blanket bogs, especially those of Ireland, take on a very characteristic aspect due to the fact that all the shallow depressions and pool-margins display an extensive although thin sward of the fine, pale green shoots and strikingly white flowering heads of the beaked sedge, *Rhynchospora alba*. This is a plant growing between about 4 and 16 in (10 and 40 cm) tall, whose lateral rhizomes are so few and short that they produce only small tufts that do not long persist: indeed the plant, reproducing freely by seed and vegeta-tive buds (bulbils), behaves rather more as an annual than a perennial.

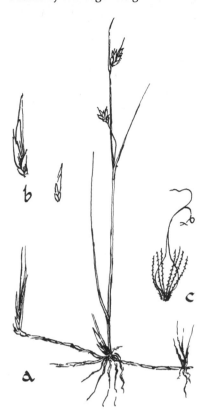

Fig. 25. White beaked sedge, *Rhynchospora alba*. *a*, rhizomatous habit of the mature plant, here shewn flowering; *b*, the turions (bulbils), which are condensed shoots that become detached and reproduce the plant vegetatively; *c*, flower with ripening biconvex fruit and its array of finely barbed bristles.

The bulbils are of two types and are formed in April to June. Those of one type are borne in the axils of the basal scales of the vegetative shoots and become detached in the autumn as the parent shoot dies down: these, given successful establishment, will produce flowering plants next season. The other bulbil type, although similar in morphology, is borne on the erect haulm of the sedge beneath its uppermost bracts: bulbils of this kind produce only non-flowering individuals in the season after dispersal. The bulbils conform to the structural type of such bodies, seen well for example in the field onion, crow garlic (*Allium vineale*), consisting of a few swollen leaf-bases around a central conical stem-apex, and surrounded by thinner and more pointed scales. These vegetative propagules suffer some damage from larvae and from birds, but play a substantial role in overwintering and dispersing the plant.

The slender flowering axes bear a few narrow leaflets and one or two spikelets containing a very small number of flowers, but these when young are snowy-white, a fact reflected not only in the Latin name for the plant but in many colloquial names for it in England and other countries. Flowering in Britain is in July and August, providing ripe fruits within a month or so. These follow the pattern of most sedge fruits

Plate *28*. White beaked sedge, *Rhynchospora alba*, flowering freely in the New Forest. (Photograph by M. C. F. Proctor.)

and are three-faced, or more often flat biconvex nutlets. These latter are especially recognisable in a sub-fossil state for the two transparent valves spring open at the base whilst still held at the beaked apex, the marginal thickening defines a clear obovate shape and very often the fruit still retains its complement of nine to thirteen basal stiff bristles, mostly armed with backwards-pointing barbs. It seems certain that the bristles aid seed-dispersal, especially perhaps by adherence to the pelt of small animals, but wind and water are probably the most effective agencies.

The white beaked sedge is found in a fairly wide range of habitats on acid peaty soils, but it remains quite typical of ombrogenous bogs and on them it occupies a very distinctive situation in regard to the prevalent water table. On this account it has become an indicator of the stage in the regeneration cycle at which the submerged *Sphagna* of the pools have just emerged to make a flat muddy surface, now about to be colonised by the large hummock-building *Sphagna* such as *S. papillosum*. Along with the abundant *Rhynchospora* there may be persistent cotton-grass, *Eriophorum angustifolium*, often the two species of sundew most tolerant of wetness, *Drosera anglica* and *D. rotundifolia*, and the earliest individuals of bog shrubs like the ling, cross-leaved heath and bog andromeda, which with bog-asphodel are all more fully represented in the next succeeding stage.

It is quite conformable with this ecological habit that the plant should be so often found sub-fossil in aquatic *Sphagnum* peat and especially in the wet stages of recurrence surfaces: it will often be found here reinforcing the evidence provided by the denizen of the deeper water, the *Scheuchzeria*, that so often indicates the swamping of an older and drier bog surface.

Bog-asphodel   *Narthecium ossifragum*

The most vivid splashes of colour on the dark expanses of the blanket

Plate *29*. The bog-asphodel, *Narthecium ossifragum*, in flower and fruit. The six stamens of the open yellow flowers are densely clad with silky white hairs up to the brilliantly red anthers. The winged seeds are escaping from the split capsules. (Photograph by K. G. Richman.)

Fig. 26. Bog-asphodel, *Narthecium ossifragum*. A sketch to shew the habit of the plant (*b*) and one of the ripe seeds with the seed-coat prolonged into two papery tails (*a*). Note the sword-like leaves arranged in two ranks and flattened in the same plane as one another.

bogs and raised bogs are supplied by colonies of the yellow-flowered bog-asphodel (*Narthecium ossifragum*), blossoming during June, July and August. The rigidly upright inflorescences, some 4 to 12 in (10 to 30 cm) tall, bear a dense cluster of star-like flowers whose six pointed petals, green outside, are a clear bright yellow within, making an admirable contrast to the bright brick-red of the six stamens, each basally clothed with fine hairs. One is slightly surprised to find that the flowers, though scented, are without nectar: all the same they are much visited by insects and for the most part pollinated by this agency. *Narthecium* belongs to the Liliaceae and by the oddest coincidence its name is an anagram of *Anthericum*, the mountain asphodels of central Europe.

Narthecium ossifragum occupies a rather well-defined level in relation to the sequence of the cycle from pool to hummock that is so typical of raised bog morphology. Its mat of narrow branched rhizomes forms a colony of upright shoots very close to mean water-level of the more persistent pools, just where the hummock-building *Sphagna* take over and where *Eriophorum angustifolium*, the many-headed cotton-grass, is replaced by *E. vaginatum*, the sheathed cotton-grass. Thus the bog-asphodel tends to form a marginal zone to the deeper pools though it does not invade the open water. Each upright shoot has two opposite ranks of upright glaucous leaves, stiff and slightly curved and admirably sword-shaped (ensiform). The plant substantially multiplies by the activity of its rhizome system, that also bears small subterranean buds which carry the plant through the winter.

The flowers give place to a beaked capsule, that splits to discharge the numerous small seeds each prolonged by outgrowth of the seed-coat into a membranous 'tail' at either end, a shape we may fairly suppose to enhance dispersal by wind, water and even rain-splash. Be that as it may, reproduction by seed in natural conditions is infrequent.

The bog-asphodel tends to grow best where there is some water movement in the soil: it has a wide range of tolerance of soil acidity and responds favourably to good mineral supply. These qualities no doubt are associated with its abundance on flush bogs, on valley bogs receiving drainage from mineral soils, and on wet places in heaths. It is said too that the plant's survival on high-level ombrogenous peat deposits is closely dependent on the mineral content of the rain water, mostly the residue of far-carried sea-spray. It might be thought that a plant of this kind should prosper in calcareous fen, but it is far too susceptible to competition from the taller fen vegetation, especially the invading fen shrubs. Accordingly we judge that to some extent *Narthecium* characterises the bogs because it finds there the openness and the sustained range of water-level that it needs, whilst tolerating the acidity and low-grade nutrition that exclude a big range of potential competitors. Its trivial name *ossifragum* recalls the colloquial German name for bog-asphodel, *Beinbruch*, the bone-breaker, a term reflecting belief in the liability to fracture of the bones of cattle that feed where the plant flourishes: certainly one would not expect high calcium in the graze of such beasts. In Britain it seems only to be lightly eaten by sheep and rabbits: better food may be too easily at hand.

The overall European distribution of the species is north-temperate and strongly atlantic, matching roughly the area of prevalent ombrogenous bog, and the sub-fossil remains, pollen and rhizomes, are found in the peat of such bogs from the time when they began widespread development towards the end of the Boreal period, seven or eight thousand years ago.

Scheuchzeria palustris

It may appear a little odd to include here a plant so rare that today it grows at only one locality in the British Isles, and that has no common English name. However, *Scheuchzeria palustris* earns its place easily enough on account of the abundance of its remains in peat bogs and by its value as indicator of past conditions.

Plate 30. Shallow pool on raised bog of Timmerhultsmosse, central Sweden, with conspicuous fringing belt of tall *Scheuchzeria palustris*, whose papery rhizomes are floating below and within the *Sphagnum* mat.

It most typically forms a submerged floating mattress round the margins of the deeper pools of the regeneration complex of undrained raised bogs, sending out its slender rhizomes, clad in papery leaf-bases, towards the open water, much as *Phragmites* does on a more massive scale in eutrophic waters. Towards the bank the *Scheuchzeria* mat is denser and supports rush-like green aerial shoots about 12 in (30 cm) tall. Many such shoots bear a terminal cluster of a few dehiscent one- or two-seeded fruits.

Although at the time I had not seen the living *Scheuchzeria*, it was not difficult for me to recognise its remains when they were first encountered in the Somerset peat diggings near Westhay. They there behaved as in the Danish peat bogs, the leafy rhizomes making a distinctive pale papery layer at a major flooding horizon, sitting on the surface of a much drier peat type. When, later on, I was demonstrating this discovery to Frank Mitchell, his keen observation enabled him also to pick out the fossil fruits and seeds: later still it has been practicable, though not very common, also to recognise the fossil pollen of *Scheuchzeria*, that occurs in the unusual form of 'dyads' instead of the commoner single grains or tetrads of most flowering plants.

I recall the pleasure of being able to demonstrate the Somerset identification to my Cambridge class, and the immediate consequence that two of the students forthwith reported a similar occurrence in the peat cuttings at Risely Moss, Lancashire. It was not long afterwards that *Scheuchzeria* was reported from several more sites of cuttings in British raised bogs, always associated with aquatic *Sphagna* and such sedges as *Carex limosa* and *C. chordorrhiza*. Although older records exist, the bulk refer to the last 2000 years, the Sub-atlantic period, when it may be supposed that the climatic turn towards wetter and colder conditions tended to induce widespread flooding of the bog surfaces. Despite the prevalence and relatively recent age of these occurrences, the living plant no longer survives at all in England and Wales, the consequence one supposes of the widespread drainage and destruction of its bog habitats, and in Scotland is found only on Rannoch Moor.

Black bog rush   *Schoenus nigricans*

The black bog rush is a densely tufted plant with smooth unbranched flowering stems from 6 in to 2 ft tall (15–60 cm), each bearing a small, compact black inflorescence that consists of a small number of closely set spikelets within a single bract and never expanding widely. The rush-like leaves are only about half the length of the haulms, and both together confer an impression of a stiff upright 'pile' on the carpet of bog vegetation, especially in the flat blanket bog surfaces of the west of Ireland where the species is especially abundant. Why it should special-ly favour these western oligotrophic mires is not really understood, but it is conjectured that it may be associated with high deposition of air-borne salt brought in by sea-spray from the Atlantic coast. This explanation is consonant with the strong tendency of the bog rush to

Plate *31*. Pool margin on raised bog in central Sweden. In the deeper water *Scheuchzeria palustris* with its clustered, dry, dehiscent fruits. In shallower water, on left of picture, *Rhynchospora alba* flowering.

Plate 32. Black bog rush, *Schoenus nigricans*, on Hartland Moor, Dorset. (Photograph by M. C. F. Proctor, May 1980.)

flourish in mesotrophic habitats or even in calcareous fen where absence of competition allows: the plant indeed sometimes occurs in coastal salt-marsh.

The tussocks of the black bog rush constitute a tight mass of packed stems surrounded by shiny leaf-bases that are characteristically dark reddish-brown to black in colour, and that make the plant easily recognisable as a macrofossil. Most of the fossil records are from Irish peat deposits and in the later part of the Flandrian period, the time when the blanket bogs shewed their major extension.

Many-headed cotton-grass   *Eriophorum angustifolium*

The genus *Eriophorum* takes its Latin name, as well as its English one, cotton-grass or cotton-sedge, from the fact that its fruits bear numerous long and silky bristles, that are in fact a modified form of perianth much elongated after flowering has been completed. These cottony hairs make the fruiting heads very conspicuous, especially in the case of *E. angustifolium*, formerly known as *E. polystachion*, in which the fruiting haulms bear several drooping heads together.

This species has a very strongly rhizomatous habit, and by these slender stems it swiftly invades the shallow water of bog pools, and retains considerable abundance in all wet bogs and acid fens, producing widespread individual aerial shoots, the leaves of which are easily recognisable in the field from the fact that a deep median channel runs down the lower half whilst the upper half is triquetrous, i.e. of triangular section, running up to an acute apex. One finds these rhizomes freely in sub-fossil state in peat, a diagnosis confirmed by the fact that they bear typically fibrous leaf-sheaths, and are associated with abundant clear pink unbranched aerating roots; the trigonous fruits are also easily recognisable.

Flowering generally takes place in May and June, followed by fruiting through June and July: of course there are always a few lingering late-fruiting heads but in mid-summer the pools and abandoned peat cuttings of the bogs make a very conspicuous display.

Plate 33. Many-headed cotton-grass, *Eriophorum angustifolium*. The multiple fruiting heads are apparent, as also the non-tufted habit. (Photograph by W. H. Palmer.)

Plate 34. Sheathed
cotton-grass, *Eriophorum
vaginatum*, in fruit all over
the bog that mantles the
glacial till below
Ingleborough.
(Photograph by M. C. F.
Proctor.)

The many-headed cotton-grass is incidentally extremely tolerant of
pollution by chalybeate water, so that it often appears to be almost the
only flowering plant present in habitats such as shallow settling pools in
the Coal Measure areas or swamps in the Greensand landscape where
the isolated tufts of *Eriophorum* grow through the scum of iron hydrox-
ides and even seem to be embedded in the thin sheets of incipient bog
iron-ore.

This cotton-grass is naturally characteristic of the early pool stages of
the regeneration cycle of raised bogs and its ecological preferences are
well suited by the general flooding associated with the recurrence
surfaces, as its abundant remains at such levels testify. The present-day
scarcity of this cotton-grass in many parts of Britain, especially the south
and east, is probably attributable to extensive drainage and peat cutting:
it is locally abundant throughout northern and western parts of the
British Isles and is in fact a plant of widespread arctic and sub-arctic
distribution, extending in Greenland as far as 83°N latitude.

Sheathed cotton-grass   *Eriophorum vaginatum*

We have stressed the prevalence of the sedges and related monocotyledons in and around the pools of the acid mires, but no sites on the bogs

are really dry and other plants of the same families also play an important role in hummock building alongside the *Sphagna* and low ericoid shrubs. In every sense outstanding in this category is the sheathed cotton-grass or cotton-sedge, *Eriophorum vaginatum*, for its large tussocks, up to a foot high (30 cm) and quite as broad, are the distinctive feature of vast acreages of bog land, both lowland and upland, where it shares dominance with such plants as the ling, deer grass or crowberry. Furthermore, in considerable upland stretches, particularly on the Pennines, this cotton-grass is now the sole dominant, its former partners having succumbed to a regime of burning, grazing and industrial pollution. I recall that Osvald, upon his first excursion to the South Pennine, '*Eriophorum*-moors' was staggered to find that his pale-grey flannel trousers (quite suitable for a Swedish trip) were soon generously striped with black soot collected from the stems and leaves of the vegetation: it was a sharp reminder of the influence of the industrial north.

The dense tussocks are produced by the habit of new vertical shoots growing upwards through the persistent leaf-bases of existing shoots and by the extreme durability of the dead leaf-bases and roots alike. The proportion of living roots to the residual dead roots in the tussock has been estimated as only one part in forty-five, and there is reason to think that individual tussocks persist for as much as a century. The stems are smooth and cylindrical below but are three-edged at the top as one approaches the strongly inflated leaf-sheath that contains the single flowering head: it is this sheathing structure, narrowed apically, that is referred to in the Latin *vaginatum*. The leaves, in contrast with those of the many-headed cotton-grass, are narrow and bristle-like, only about 1 mm wide.

Flowering in April and May is succeeded by general fruiting about a month later, although sometimes there may be a second, late flowering in the autumn. Within the spikelet one may find that the glumes are distinctly silvery to slaty blue-black in colour; and the bristles below the ovary grow as the fruit ripens to achieve a length of 2 cm, and are of a snowy whiteness that contributes to the startling display of the cotton-grass communities in mid-summer. The local name of 'hare's tail' is quite understandable. The fruits themselves are trigonous nutlets that appear to need exposure to light for successful germination.

The resistance of cotton-grass to decay is such that peat diggers everywhere objurgate the dense fibrous 'mabs' of old stem-bases that defy their spades and beckets. These of course make easily recognisable fossils and so do the black unbranched aerating roots which originally grew down into the anaerobic lower peat layers. One now usually finds

Plate 35. Sheathed cotton-grass, *Eriophorum vaginatum*. Note the single fruiting heads and the fact that most of the stems and leaves come from the large tussock. (Photograph by W. H. Palmer.)

them concertina'd by subsequent compaction of the peat and if one unfolds the zig-zags one has a measure of the degree of compression undergone. Even wooden piles will shew a similar folding in compacting peat as Bulleid realised in recording the palisading of the Glastonbury Lake Village. Since the sheathed cotton-grass responds favourably to improved drainage, as upon the rand in raised bogs or beside drainage channels of any kind and as participant in the later stages of the regeneration cycle, it has long been known to have left thick layers at stages suggesting former dryness of the bog surfaces. Thus the recurrence surfaces are often represented by dense growth of cotton-grass at the top of *Sphagnum–Calluna* peat, before the onset of a flooding episode and bog regeneration.

The sheathed cotton-grass has a very wide circumpolar distribution, growing commonly on the arctic tundra of North America and Russia, so that it is not surprising that it was present in this country in Late-glacial time and presumably survived the glaciations here. The bulk of records however, naturally come from the the milder 'Post-glacial', when the development of ombrogenous mires provided opportunity first for the growth and then for the entombment of this very recognisable species.

Deer grass   *Scirpus caespitosus (Trichophorum caespitosum)*

The deer grass has been for a long time referred to the genus *Scirpus*, the

Linnaean name I still retain as a matter of convenience, although it is now placed in *Trichophorum*. The vast proportion of its British population belongs to the sub-species *germanicum*: only in a few isolated mountain localities are there populations of a second sub-species. For the purposes of this account we need only think of the species in the broad sense, i.e. *Scirpus caespitosus* agg.

The growing plant is made of stiff slender shoots lacking any broadened leaves. They build tough clumps of radiating straight stems, each enclosed basally by dark, rich brown shining leaf-sheaths. Such shoots are in fact sympodial, remaining vegetative and about 4 to 6 in (10 to 15 cm) high in the first year, continuing in the second by the development of a new lateral stem that bears a solitary small spikelet of flowers. As in so many monocotyledons, the shoot at this stage is really a composite structure of cylindrical leaf-bases (sheaths) like a half-extended telescope from within which the stem proper, with its inflorescence, eventually emerges. The uppermost leaf-sheath bears a tiny terminal projection that we can regard as the vestigial leaf-blade. The green shoot is smooth and cylindrical and the solitary spikelet at its apex is only an eighth to a quarter of an inch (3 to 6 mm) long, with very few flowers. These, which are hermaphrodite, produce small obovoid, three-angled, shiny fruits surrounded basally by half a dozen short brownish bristles.

Flowering takes place in June and July, followed a month or so

Plate 36. A flowering tuft of deer grass, *Scirpus caespitosus*, growing in an unusual habitat, the wet cliff-face of ancient volcanic rocks on Cader Idris, central Wales. (June 1957)

Fig. 27. Deer grass, *Scirpus caespitosus*. Sketch of a small plant that already displays the tufted habit. Note the basal, glossy concave scales, dark in colour, that long remain recognisable in peat.

afterwards by production of ripe fruit. Neither of these stages makes the plant conspicuous and in general its uniform bristly shoots confer the appearance of very rough sward, yellow for much of the year, over large stretches of upland peat bog. The deer grass is too short and too tough to be susceptible to grazing and this no doubt assists its prevalence: it is also generally free from insect depredation.

Scirpus caespitosus is common not only on ombrogenous bogs but also on peaty soils with some mineral content, such as often bear heath in sites contiguous with acidic bogs degraded by cutting and burning. On raised bogs the deer grass is widespread, entering the regeneration cycle at the stage where the hummock builds up above water-level, and persisting in the mature stage, sometimes, as for example in the big western bog at Tregaron, forming a *Scirpetum* where it dominates the closely set hummocks that carry in lesser amount also the ling, cross-leaved heath and the two cotton-grasses. It is, however, on the high-altitude blanket bog in regions of heavy rainfall, such as western and north-western Scotland, that the deer grass exists extensively as a vegetational dominant, occupying in fact regions and areas very similar to those dominated by cotton-grass. Similar *Scirpeta* occur locally on Exmoor, Bodmin Moor and the Wicklow Mountains. On other lower-level blanket bog areas, such as those of western Ireland, the plant is an important, though not dominant component, taking second place to *Schoenus, Rhynchospora* and *Molinia*. In peat bogs such as these deer grass is itself an important peat former and in peat cuttings its remains are easily recognised by the dense tussocks still clothed in dark shiny scales. As in so many similar bog species, the fossil record largely reflects the increased development of ombrogenous bogs in the second half of the Flandrian period.

Purple moor grass *Molinia caerulea*

A last member of the category of tussock-forming bog plants is *Molinia caerulea*, the purple (or blue) moor grass, which is co-dominant or abundant over great stretches of blanket bog and is locally dominant on raised bogs, fens and wet situations on many kinds of mineral soil tending to acidity and infertility.

It has a very short rhizome indeed, that may, where water-level is consistently high, as in some blanket bogs, make a horizontal branching system from which the vertical green shoots arise. It is far commoner, however, that these stubby rhizomes aggregate into a tough perennial root-stock, forming a tussock that may be as much as 3 ft (1 m) across and half this in height. The tussock is anchored by numerous long,

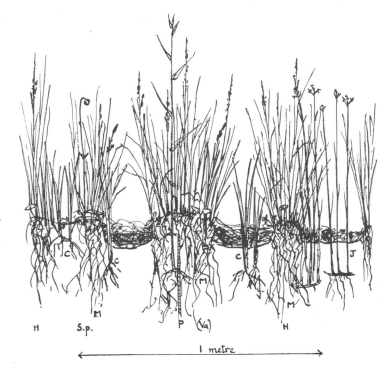

Fig. 28. Purple moor grass, *Molinia caerulea*. Transect diagram through annually mown litter at Wicken Fen, Cambridgeshire, to show the strongly tufted habit of the grass. It is accompanied by fen species: C, *Carex panicea; J, Juncus obtusiflorus; P, Phragmites communis*, etc.

sharply twisted, yellow 'cord' roots of quite remarkable strength. The leading buds of the rhizome segment form the aerial shoots of the current year. The finely tapered leaves are delicate in texture, soft and more or less pilose, often bluish-green in colour and distinctive in that they do not bear a ligule at the base of the leaf-blade, but a ring of fine hairs. The flowering shoot may be over 3 ft (1 m) tall, and the panicle or flowering head itself is perhaps a third of this. It is green to purplish in colour, or sometimes slaty blue, a character contributing to the colour adjectives in the Latin and colloquial names, and it has a 'strict' habit, that is to say not spreading widely and narrow in general outline.

As the season progresses, nutrients are passed to the basal internode of each aerial shoot, to be stored as starch and hemicellulose. The tuberised stem-base, swollen and hard, at the season's end sheds the aerial stem by an abscission layer that leaves a circular scar at the top of a body about 1 in (2.5 cm) in length, and shaped like an Indian club. Next year's shoots arise from fresh buds on the rhizome, drawing upon the persistent 'bulbils' for food: these may persist for two or three seasons and afford an unmistakable means of recognising the *Molinia*, living or fossil.

The stem-tuber formation and annual shedding of the shoot are at the base of several aspects of the ecology of the purple moor grass. Because

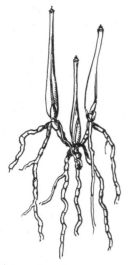

Fig. 29. Purple moor grass, *Molinia caerulea*. A small portion pulled away from the side of a tussock in winter condition, shewing three of the hard stem-tubers; two will give rise to shoots next year, the third was exhausted by making a shoot last year. All three shew the scars where the green stem and leaves abscissed. The tubers arise from a short stem that also carries numerous yellow contractile 'cord' roots of great strength.

the shoots are deciduous and soon decay, it is only for a short period in summer that they form a complete ground cover: on this account many dwarfer species with an earlier vegetative season flourish along with it, such (in fenland) as marsh pennywort, carnation sedge and marsh valerian. Autumn crop-taking does not harm *Molinia* since nutrients have been withdrawn into the basal tubers that lie below scythe level, whilst such cropping severely damages and soon kills a tall evergreen sedge such as *Cladium mariscus*, thereby converting a 'mixed-sedge' *Cladio-molinietum* into the more open 'litter' community, a *Molinietum*.

Annual cutting of *Molinieta* for litter was formerly practised in the English Fenland, as it is widely in continental Europe. No doubt the tuberisation and close tussock formation are partly responsible for the considerable resistance of the plant to burning, which may thus be favoured by the repeated burning of moors.

A tussock *Molinietum* is one of the most awkward of plant communities to traverse. The hummocks are so large and steep-sided that it is very difficult to overstep them, the clefts between, concealed by overhanging foliage, are dangerously narrow, and it is impracticable to walk on the rounded tussocks themselves.

The general consensus among ecologists is that *Molinia*, whilst requiring more or less continuous wetness in the soil, needs the soil water to be moving and avoids situations of total water-logging. These limitations are constantly apparent in local field situations. On most raised bogs *Molinia* is associated with the tussock-forming stage of the regeneration cycle, but it is far more vigorously developed on the sloping rand, on the banks of drainage channels and round the occasional swallow-hole, all situations in which the drainage has been sharpened. Sometimes, as in Tregaron bog, an area of the bog surface may bear a tussocky *Molinietum* perhaps in response to recent climatic change or human interference. We have already referred to the abundance of *Molinia* in the western Irish blanket bog along with ling, beaked sedge, bog myrtle, cross-leaved heath, deer grass and bog rush. On the uplands of northern and western Britain wet blanket peat bearing *Molinia, Sphagna* and bog myrtle seems to represent areas that have been recently converted from birch woodland to bog, and it is common to find *Molinia* grassland at the lower margin of cut or eroded peat bogs where they are fed by flushing. It seems likely that the basal peat of the Pennine blanket bogs likewise records succession represented by birch woods with *Molinia* being invaded by *Sphagna*. Compared with the other abundant plants of high-altitude bogs – deer grass, cotton-grass, moor rush, ling and bilberry – *Molinia* is found on analysis to have a relatively high mineral content. None the less the absolute amounts are small and it is a poor fodder plant, of value for grazing mainly in the period of spring growth. What strikes one with special force is the extremely low concentration of soil nutrients the moor grass will tolerate whilst at the same time it can flourish in highly calcareous fen peat, in which its roots may indeed penetrate layers of shell marl that are very largely calcium carbonate. The plant moreover shares with *Eriophorum angustifolium* and one or two other sedges and the rush, *Juncus sylvaticus*, a remarkable toleration for iron, so that it often marks the site of flushes of iron-rich water.

Although the preference of living *Molinia* for some aeration leads to its remains tending to decompose during peat formation, the twisted cord roots and flask-shaped tubers remain long recognisable, providing us with clear evidence of its presence on our mires at least from the onset of the Atlantic period.

7

Plants of the bogs: dwarf shrubs etc.

Ling (heather)   *Calluna vulgaris*

Well known and well loved as the ling (*Calluna vulgaris*) may be, the plant always referred to in England as the 'heather' deserves at least some comment as a species characteristic of acid mires, for no plants save the *Sphagnum* mosses occur so abundantly or so consistently in these habitats. It is of course true that the ling also occupies a far wider range of situations and soils, extending to slopes and exposures that could not carry bog, and that it constitutes the dominant in heath lands, upon dry acidic soils, where both grazing and burning greatly affect its performance.

The prevalence of ling on the bog surfaces is matched by the frequency of its sub-fossil remains in the post-glacial peat. The crooked stems with wrinkled bark strongly resist decay and frequently still are found bearing sprays of young shoots clothed with densely ranked leaves, often so close together as to be 'imbricate' and exhibiting not only the typical tetragonal cross-section but the basal prolongation into two spurs partly clasping the supporting stem. In addition well-preserved flowers commonly occur and less often the capsules and the minute seeds. In recent years moreover pollen analysts have regularly identified the *Calluna* pollen, so that it might well be claimed that no species had a more fully documented post-glacial history than this.

It is a dwarf woody shrub that in open habitats generally does not exceed $2\frac{1}{2}$ ft (0.8 m) in height, and that roots freely from prostrate branches when growing in moist conditions. It is evergreen, bearing leaves that individually may last for three years and it does not have protected winter-beds. These qualities are clearly associated with the preference of the plant for cool oceanic or sub-oceanic conditions and a pronounced atlantic distribution range, very apparent on the European map of its occurrences. It would seem also that these climatic preferences (even requirements) are associated with the abundance of *Calluna* on both raised bog and blanket bog, for it is distinctly susceptible to

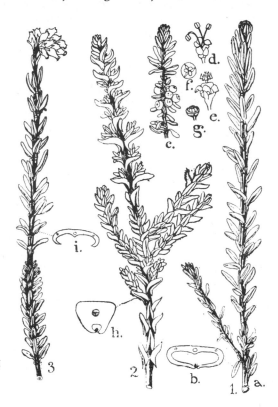

Fig. 30. Dwarf shrubs of acid peat bogs. 1 *a–g*, crowberry (*Empetrum nigrum*); 2 *h*, ling (*Calluna vulgaris*); 3 *i*, cross-leaved heath (*Erica tetralix*), to shew arrangement, shape and cross-section of the leaves. For the crowberry there are sketches also of flower and fruit. (Drawing from T. W. Woodhead, 1917.)

summer drought, a limitation no doubt associated with a very shallow rooting habit which is itself tied to intolerance of water-logging, the fine roots growing only in the upper aerated levels of the soil. Thus, despite its ubiquity on acid mires, the ling tends to be restricted to the better-drained situations upon them, such as the tops of hummocks, and to show most vigorous growth at the crest of the rand or on the edges of drainage cuts and peat diggings. It is not surprising therefore that *Calluna* should be abundant in the *Stillstand* stage of mature bogs, or that dense layers of its remains, along with those of *Eriophorum* (cotton-grass) should constitute the dry phase of regeneration surfaces of raised bogs before the ensuing swamping.

In the phase of botanical thought, now passing, of 'argument from design', it was usual to seek to explain the small surface of the leaves, their lower stomatal surfaces enclosed by revolute margins, close packing and heavy cuticles, as devices meeting a need to cut down water loss. Such 'xeromorphic' adaptations seem certainly inappropriate in a plant so typical of bogs and ingenious hypotheses were put forward that the bog water was somehow unavailable to the root-systems. These hypotheses and the adaptive arguments together still remain unsub-

Plate 37. Ling, *Calluna vulgaris*. Large bush in full flower on raised bog at Fenn's Moss. The fallen flowers that here strew the bog surface, are often found in peat preparations.

stantiated for *Calluna* as for the many other genera and species of the Ericaceae common on our mires.

It is of interest that it was in *Calluna* that much of the pioneer research was done that demonstrated how in many plants there is a regular symbiotic relationship between the active root-system and the microscopic strands of a living fungus that in some cases lives in the outer cells of the living root and in others clothes the rootlets with a fine mantle. In either instance this 'mycorrhiza' replaces the normal surface of fine root-hairs through which absorption of water and dissolved nutrients takes place. It soon was apparent that this relationship was general and obligatory in many families of the plant-kingdom, including all the Ericaceae, and a vast amount of research has been done, and continues, upon its significance and the mechanisms by which the fungus, the higher plant and the soil interact with one another.

The ling is easily distinguished from the plants of the genus *Erica*, the bell-heathers, by the flowers, of which both calyx and corolla have fully separate components, the four sepals similar to but larger than the four petals. The flowers produce copious nectar, sit horizontally on the flowering shoots and are visited by a host of pollinating insects, prominently among them the bees.

The dry capsules that succeed the flowers produce vast quantities of dusty seeds that the wind will disperse to a considerable distance and that germinate in suitably moist conditions so that regeneration by this means is common.

Whereas many of the dwarf bog shrubs, like the crowberry, cloud-berry and bilberry, are ecologically linked to the population of northern birds by the attraction of edible fruits, in the ling (as indeed in crowberry also) it is the cropping of the tender young shoots by the grouse that establishes a similar dependence. Upon this in turn there hangs the whole tradition of maintaining grouse moors and keeping them in high productivity by regular and considered firing. The heather also serves importantly for browse of mountain sheep and cattle. It is not hard to see how cropping and burning together are major ecological factors in the variable and extensive range of communities dominated by 'the bonnie purple heather'.

Finally we may note something of the role played by *Calluna* in promoting the spread of ombrogenous mires on soils originally bearing a woodland cover. The humic acid complexes produced by decay of the heather combine readily with iron in the mineral soil to form soluble compounds that, where rainfall and soil texture are favourable, are easily leached down to lower layers. Thus there is encouraged 'podso-lisation', that results in an upper or 'A' horizon of the soil, strongly leached of plant nutrients and white in colour from loss of its iron-oxide matrix, above a 'B' horizon, dark brown, red or black in colour from re-deposition of the down-leached iron and humus. It has been re-peatedly shewn in Denmark and Britain especially, that where prehis-toric man, from Neolithic time onward, cleared deciduous woodland on sandy soils, the subsequent entry of *Calluna* led to this podsolisation where the acidity and poverty of the 'A' layer and the impermeability of the 'B' layer together prevented tree-regeneration and provided condi-tions for growth of acidic heath communities, or even, when climate and locality favoured this, development of ombrogenous mire. Such vegeta-tional shift was no doubt greatly favoured during the Sub-atlantic with its cool moist climate and Iron Age agriculturalists bringing iron tools to enthusiastic woodland clearances.

Cross-leaved heath *Erica tetralix*

The ubiquity of the ling on our acidic mires is approached by that of its relative, *Erica tetralix*, the cross-leaved heath, identifiable readily enough by its leaves, typically borne four in a whorl, and very strongly recurved over the lower surfaces. Less close inspection is needed to recognise the terminal heads of rosy-pink flowers, with their four petals united into rounded dropping bells that contrast attractively with the general grey-green pubescence of the whole plant.

Even more than *Calluna*, *Erica tetralix* is a strongly oceanic plant, its

Plate *38*. Cross-leaved heath, *Erica tetralix*, in flower in central Wales. (1957)

range strictly limited to the western areas of Europe, and within Ireland, England and Wales displaying a much more general occurrence in the west: in Scotland it is almost ubiquitous and forms along with the ling a consistent component of many important communities of wet heaths and mires, associated respectively with crowberry, purple moor grass and the deer grass (*Scirpus caespitosus*). On the blanket bog of western Ireland it is consistently associated with the black bog rush, *Schoenus nigricans*, that has very similar growth requirements.

Although so strongly associated with the ling, the cross-leaved heath is on the one side less tolerant of drought and on the other more tolerant of water-logging: it is also commoner in flush mires, where there is increased base status and some improvement in aeration. Abundant as *E. tetralix* is on all oligotrophic mires, it also grows widely on acidic heath if this is sufficiently wet.

Crowberry *Empetrum nigrum*

Blanket bogs in the British Isles, especially those of the north and west, very often support considerable growth of the crowberry (*Empetrum nigrum*), a dwarf, straggling evergreen shrub with dark green leaves borne densely upon upright shoots typically red in colour. It is a plant seldom taller than 20 in (50 cm), and it often adopts, especially in exposed, wind-swept places, a very prostrate habit, sending down adventitious roots very freely so as to establish a large creeping mattress:

indeed the plant persists in this way much more than through seedling establishment.

The species is dioecious, that is to say, that as in the common nettle, bryony and dog's mercury, there are separate male and female plants. Naturally the flowers in the two sexes differ in appearance and in the crowberry the pink wind-pollinated ovulate flowers, though quite small (1 to 2 mm across) are nevertheless larger than the staminate ones. The female plants carry a crop of fleshy black fruits that are eaten by various animals but especially by the grouse, which bird is probably the main agent of seed dispersal. Again it is mainly the grouse, along with ptarmigan, that chiefly browses the young green shoots.

Although the crowberry is a highly characteristic plant of ombrogenic mires its range extends to the acidic peaty soils of moors and heaths and it even occurs sometimes on such calcareous soils as the Sugar Limestone of the North Pennines and the Carboniferous Limestone of the Burren in Western Ireland.

There seems no doubt that the crowberry favours a cool moist climate, and is correspondingly far commoner in northern and western Britain than in the south and east, although there is good fossil evidence that at the transition from the Glacial to Post-glacial time it was widespread and abundant, so that some investigators have written of '*Empetrum* heaths' for that phase of our vegetational history. Despite the general preference of *Empetrum* for distinctly wet soils, none the less on the bogs themselves it favours the drier habitats, such for instance as the crowns of large hummocks, the drier margins of raised bogs and the edges of the peat hags in blanket bog subject to gullying. In the last situation particularly its preferences seem closely shared by the bilberry (*Vaccinium myrtillus*) and the cloudberry (*Rubus chamaemorus*).

It is in good general conformity with the attributes we have mentioned that crowberry should be able to attain an altitudinal range of about 3000 ft (920 m) in North Wales and Northern England: it is harder to assess its response in Scotland for there *E. nigrum* appears to be very largely replaced by the tetraploid *E. hermaphroditum*, a species with double the chromosome complement of *E. nigrum* and with flowers that bear both stamens and ovaries.

The wedge-shaped 'stones' from the crowberry fruits are long-lasting and recognisable in the fossil state, but it is the thick-walled pollen tetrads that have been an especially useful index to the plant's past distribution and abundance, occurring as they do in such numbers as to allow the construction of range and frequency maps for successive periods of the Quaternary.

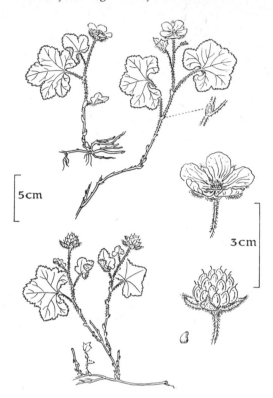

Fig. 31. Cloudberry, *Rubus chamaemorus*, shewing the habit, flowers (white) and fruit (orange). (From Roles, 1960.)

Cloudberry *Rubus chamaemorus*

The botanist of southern Britain is liable to be quite unfamiliar with the very distinctive and attractive cloudberry, *Rubus chamaemorus*, that is so characteristic of high-level blanket bogs dominated by cotton-grass and ling, prevalent in the Pennines and the Scottish Highlands. This northern or even sub-arctic plant is unlike most plants of the genus – blackberry, raspberry and dewberry for instance – in consisting of a much-branched, woody system of slender underground stems that give rise, during the summer, to short upright vegetative shoots each with just a few leaves and perhaps a solitary flower. The leaves are divided into five to seven lobes and are rugose with very projecting veins. As with the crowberry, cloudberry is dioecious, though both male and female flowers are large, white and are visited by the pollinating agents, mostly flies.

Each female flower produces a cluster of druplets in the manner of the genus, but in the cloudberry these number between four and twenty and in ripening turn from red to a clouded apricot hue. These fruits have an exquisite flavour best appreciated after hot mid-summer field work on the high northern bogs. A visitor to Swedish Lappland at this season

Plate 39. Cloudberry, *Rubus chamaemorus*, on an Aberdeenshire bog, shewing the coarsely serrated palmate leaves and conspicuous white flowers. (Photograph by M. C. F. Proctor.)

may see the opening of the fruit-picking season when, at a given signal, a line of Lapps advances across the flat bog surfaces, rapidly filling the birch panniers with fruit destined to travel 1000 miles south by train to Stockholm. The quick-frozen fruit, or still more, the liqueur made from them, are rightly esteemed delicacies. *Au naturel*, and especially in Britain, the bulk of the fruit is taken by the red grouse that thus serves to spread the plant. None the less, the plant mainly extends vegetatively by its rhizomes and these provide for perennation by dormant buds situated at or just below soil-level.

Although the cloudberry is strongly (though not exclusively) associated with ombrogenous peat bogs, it does not tolerate water-logging and is generally more abundant or better grown where the drainage is sharpest, as on bog margins, on tussocks and beside drainage channels. Perhaps this is why it is so seldom found on bogs dominated by the deer grass.

Both cloudberry and crowberry are very northern in their range and in Britain it is mostly at high altitudes that they find the cool moist conditions that they need. Accordingly they enjoy only a short growing season and vegetative growth is arrested throughout the long winter. In crowberry adjustment to these conditions appears in the way in which the terminal shoot ceases growth in September or October, persisting as a densely wrapped bud that stays dormant until about April. In the

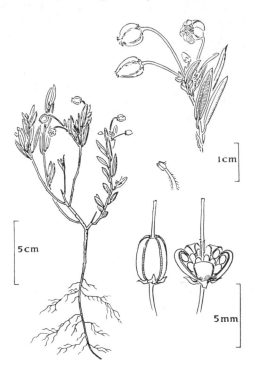

Fig. 32. Bog andromeda, *Andromeda polifolia*, shewing habit, narrow leaves with reflexed margins, apically clustered flowers and capsular fruits. (From Roles, 1960.)

cloudberry it is the rhizomes that develop dormant buds just below soil-level in the autumn and then resume activity in May. Both species shew their adjustment to the cold environment in the necessity for low temperatures to break winter dormancy, of the seed in *Empetrum* and of both seed- and bud-dormancy in *Rubus chamaemorus*.

The close association of both species with bird dispersal, especially by grouse and ptarmigan, has been noted: the bilberry, *Vaccinium myritillus*, is similar in this regard, as also in favouring those parts of the bog surfaces where the local drainage tends to exclude water-logging. One cannot escape the view that in their common ecological adjustments these northern bog plants reflect strongly the character of the upland mires they so consistently inhabit.

Bog andromeda *Andromeda polifolia*

The Latin name, *Andromeda*, chosen by Linnaeus for this genus of handsome plants, was that of a mythical princess whose beauty was said to rival that of Juno herself. The one British species, *A. polifolia*, seems to have no genuine colloquial English name, and only a north German name of bog-rosemary recalls the plant itself and its typical foliage. It is a low evergreen shrub with creeping horizontal stems running within the *Sphagnum* moss carpet, and giving rise to sparse erect shoots, some 4 to 12 in (10 to 30 cm) tall, that are sparsely and narrowly branched and bear the foliage and flowers. The highly typical evergreen leaves are dark green above with margins that are strongly revolute to the blue-green lower surface; they are narrowly elliptic in shape with entire margins and reach a length of $\frac{1}{2}$ to $1\frac{1}{2}$ in (1.5 to 3.5 cm).

Handsome as the living plant is, the species first caught my own attention when I began to note the frequent and unmistakable display of the shoots and leaves spread horizontally in sheets of aquatic *Sphagna* at particular horizons in the dried-out peat bogs of the Somerset Levels. No other British plant had revolute leaves of this size and shape, so that its identity was certain although Somerset is outside the present range of *Andromeda*. The fossil evidence itself quickly showed that these layers in the peat were always part of a recurrence surface, representing the onset of a general flooding of an older and drier bog surface. This was confirmed in many subsequent field observations by ecologists who came to associate the bog andromeda especially with the carpets or 'lawns' of *Sphagnum* formed during flooding episodes, or with the margins of inundated tussocks.

The flowering shoots carry a modest number of flowers, rosy-pink and each pendant from a long curved pedicel. The urn-shaped corolla encloses ten stamens, hairy at the base and dehiscing by apical pores as is common in the family, and the anthers are made more likely to be

Plate *40*. Bog andromeda, *Andromeda polifolia*, shewing the terminal clusters of drooping rose-pink flowers and the leathery dark-green leaves with revolute margins. (Photograph by W. H. Palmer.)

Plate *41*. Macrofossil remains of *Andromeda* on the split surface of a sample of raised bog peat brought back by Professor West from Holme Fen, Huntingdonshire. The characteristic leaves and straight stems of the bog andromeda are highly recognisable. The finer stems and smaller leaves of the cranberry are present, as also the large ovate leaves of another *Vaccinium* species.

shaken by visiting butterflies and bees through the presence of two long horn-like extensions. The flowering time is from May to September and flowers are succeeded by sub-globose capsules of small dry seeds.

Andromeda extends in cold-temperate and arctic latitudes round the northern hemisphere, but in the British Isles it seems to be restricted to central latitudes, not now growing north of west Perthshire or south of a line from Cardigan Bay to the Humber, whilst in Ireland it is primarily found in the Central Plain. It seems possible that this is because of the plant's very strong, almost exclusive association with raised bogs. The great destruction of these mires by cutting and drainage in southern Britain has certainly diminished the historic range of *Andromeda* although it has considerable powers of survival in even severely damaged habitats.

Sweet gale *or* bog myrtle   *Myrica gale*

When we are engaged in deciphering the past history of the peat bogs by recognising the presence and disposition of the plants that once grew upon them, we are most often concerned with those species that are so typical of acid mires that they are seldom absent, as for instance the ling,

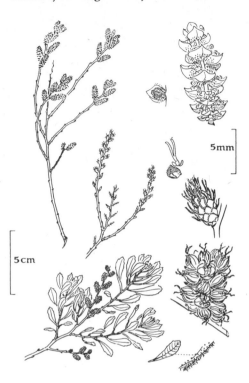

Fig. 33. Sweet gale or bog myrtle, *Myrica gale*, shewing habit, leaves with apical serration and dense glands, cones, flowers and fruit. (From Roles, 1960.)

bell-heathers, *Sphagnum* moss and cotton-grasses, and may moreover tend to grow to some extent exclusively on such bogs. Perhaps in the latter class we should place the *Scheuchzeria*, the bog andromeda and cloudberry.

It is useful all the same, sometimes to make use of a species with quite different relationship to the mires, for instance the sweet gale or bog myrtle, *Myrica gale*. This is a well-branched shrub about 3 ft (1 m) tall that spreads readily by suckers and is very readily recognised despite its very inconspicuous flowers grouped in short catkins on both the male and female plants. The leaves, borne annually, are grey-green and pubescent below, about $\frac{3}{4}$ to $2\frac{1}{2}$ in (2 to 6 cm) in length, oblanceolate with an overall wedge shape at the base like those of the sallow. They can, however, infallibly be recognised by having marginal serrations on either side of the apex, a character easy to see in the sub-fossil material. More striking, as appealing to other senses also, the leaves are richly strewn with glistening gold hair-glands that convey an aromatic intoxicating scent to the air. One uses the adjectives advisedly. Like the less common ericoid shrub *Ledum palustre*, sweet gale was formerly added to beer for its bitter flavour and possibly as a preservative. In this way it anticipated the hopping of beer, and we are asked to believe that an infusion of bog myrtle was drunk by the Vikings before entering battle and that the drink sent them berserk. Not surprisingly the powerful smell led to many folk uses, among them the English habit of putting the leafy shoots between the bed-clothes as a flea repellent.

The plant is wind-pollinated, the four stubby stamens of each flower shedding pollen into the hollow bowl of the subtending bract, whence it is dispersed by the shaking of stiffer gusts of wind. The overwintering flowers open in April and May and the carpellary flowers finally produce a dry compressed one-seeded fruit given a pair of wings by its adnate bracteoles.

Fossil recognition of the plant is easy, by means of the macrofossils, especially by the serrate leaf-apices with glands, by the fruits and by the twigs which have the stumpy bases of lateral branches growing from them closely and at right angles. The pollen is another story! It very closely resembles that of the hazel and has presented in the past a great obstacle to the accurate assessment of the spread of hazel in Britain. The trouble is the worse since different pollen analysts cannot reach agreed conclusions on how to separate pollen of the two genera, and it is a common convention to link the count of them both as 'coryloid' type.

I first encountered the sub-fossil remains of bog myrtle at the major flooding horizon in the bogs of the Somerset Levels, where all the recognisable remains mentioned occurred in a layer with abundant rhizomes of the giant sword sedge, *Cladium mariscus,* set in a matrix of *Hypnum* moss peat. It was naturally interpreted as the result of swamping of the bog surfaces by mineral-rich water, and indeed this plant assemblage still characterises the scattered fen situations, such as Wicken Fen, from which virtually all trace of raised bog has been removed, but where living relicts may persist. In north-western Europe bog myrtle occurs in 'wooded raised bogs' and 'shrub-bogs' along with the cowberry, *Vaccinium vitis-idaea,* and the bog andromeda, *Andromeda polifolia,* but these are bog types barely represented here. It occurs on bogs, wet heaths and fens, in situations shewing a tendency to mesotrophy, avoiding the open bog plain save where it occupies the summit of large hummocks; and it tends to grow in marginal situations, as for instance in the margins of the lagg where acidic fen communities prevail.

One should add that the plant seems not to tolerate persistent water-logging, perhaps an attribute associated with the fact that it bears on its roots nodules containing nitrogen-fixing bacteria. It grows more or less throughout the British Isles but there is no mistaking its strong preference for the north and west.

8

Recent peat of Somerset: a double inundation

Our investigations in the Somerset Levels centred upon the flat landscape between the Polden Hills and the Wedmore Ridge, with Glastonbury at its landward end, and with sundry low islands, particularly at Meare and Godney, protruding midway between the ridges. A large part of the country was lush pasture for dairy cattle, in fields intersected and drained by numerous artificial straight channels, the 'rhines', in which water-level was seldom more than a foot or so below field-level. It took some time fully to realise that this vast extent of fertile pasture had been produced by the cutting away of the ombrogenous bogs that formerly filled the whole extent of the valley, the more so because the water-plants in all these water channels, large and small, were evidently quite eutrophic with a full and characteristic range of fen species. This character shewed up most strikingly where active peat cutting was still extending into the region of drained, but otherwise largely intact raised bogs between the island of Meare and the northern slopes of the Poldens, the area comprising Shapwick Heath and Meare Heath.

Going along the road that runs north from Shapwick village to Westhay, one saw the level green pastures on the left, whilst on the right was the bush-grown edge of the old raised bogs, then still carrying the oak–birch wood that had colonised the bog surface after its drainage. Leading on to the bogs were the occasional peaty droves along which dried 'turf' was removed by cart or lorry to the main road, often by way of huge peat stacks left temporarily on the road verges. It was possible to see areas on the bog margin where cutting had ceased only a few years previously, and where after extraction of the peat down to the general water-level, the ground surface had been restored in accord with the turbary contract. The exceedingly soft surface of discarded peat sods was now being knitted together by the spontaneous growth of pasture and meadow plants nourished by the repeated winter flooding of calcareous water derived ultimately from the great catchment area of limestone rocks exposed on the surrounding hills. We were to find that this incursion of lime-bearing flood-water into the raised bog landscape,

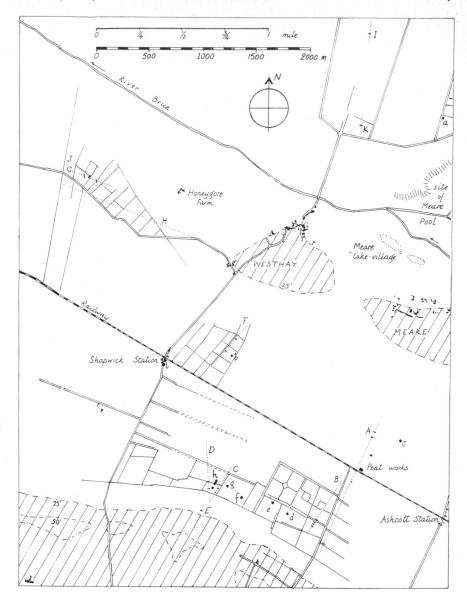

Fig. 34. Sketch-map of the Shapwick–Meare–Westhay region of the Somerset Levels, shewing what was known by 1960 of the prehistoric wooden trackways, here indicated as dotted lines labelled *A* to *J*. Since that date our knowledge has extended considerably but it is clear that the tracks link up the hill-ridges and the low islands. Archaeological discoveries are indicated by black lettered dots.

now of such dominant significance for the dairy industry of the Levels, has been a feature of great importance in the evolution of the bogs themselves throughout at least the last three millennia. In some places, rather too far from hard roads to be easily convertible to dairy pasture, the cutting away of peat to the lowest practicable level had left a rougher, though generally flat surface which had been colonised by bushes, mostly of birch, sallow and sweet gale (*Myrica gale*), with occasional plants of marsh-fern (*Thelypteris palustris*), royal fern (*Osmunda regalis*), giant water-dock (*Rumex hydrolapathum*) and milk-parsley (*Peucedanum palustre*) in less shaded spots. Such communities, bearing the impress alike of fen and acid bog, and themselves seral (succession-al), offer a great range of plant and animal life and a few isolated patches of such have been acquired as nature reserves.

In 1941 my close friend of Cambridge days, Dr A. R. Clapham, then lecturer at Oxford, joined me in a programme of intensive field work on the stratigraphy of the bogs in the Shapwick–Westhay region. We began by trying to recover the prehistoric wooden trackway that had been described by Bulleid in 1933 as running north and south across Meare Heath, probably at least 1½ miles (2.4 km) long, and of very robust construction. It consisted essentially, according to him, of transverse balks of timber, mostly oak, some 2 m or more long, laid transversely in rather irregular sequence along the line of the track.

The balks were either split half-trunks or beams, flat on both faces, and many were perforated by mortise holes, usually square-cut, still retaining the squared and sharpened vertical stakes that held the track in place. A certain amount of birch and alder brushwood strengthened the track, forming a layer between and below the main sleepers. We need not now say more of the track's construction, which was ultimately elucidated only in the 1970s, through extensive field excavation by an active group of mainly student archaeologists organised by Dr John Coles from Cambridge. Bulleid had reached no conclusion as to the age of the track and we were hopeful that the bog stratigraphy and pollen analysis might allow us to make good this deficiency. We were fortu-nately able to expose the edge of a section of Bulleid's track (Meare Heath track) left under a peat bank preserved to carry a light railway, and immediately two dramatic observations followed. Firstly, the undis-turbed track was lying upon the upper surface of the dark-brown highly humified *Sphagnum–Calluna* peat, such as we had already proved on Shapwick Heath to have an abrupt junction with an upper, paler, less humified peat. This fact alone, if European comparisons were to be trusted, gave a first indication that the trackway might be an artefact of the Late Bronze Age, a conclusion sustained by the tree-pollen diagram when we had had time to construct this from our sampling at the site. The second important stratigraphic observation was that only a short distance above the main timbers of the trackway there was a dark-brown peat containing abundant rhizomes and fruits of *Cladium mariscus*, the giant sword sedge, a plant of wet fens of moderate to high base status, with which we were already familiar since it still is the dominant plant of large areas of Wicken Fen and parts of the Norfolk Broads. Here it was accompanied by abundant *Hypnum*-type mosses and the characteristic spiky twigs and serrate leaves of the sweet gale, a plant also of fens albeit the more mesotrophic. We were able, when we returned in the following year, to find the southern end of the Meare Heath track as it approaches the foot of the Poldens, but we were also told of the existence of a much lighter trackway, made of brushwood supported by short poles and longer stakes. It was said by the peat-works foreman, Mr Foster, to be recognisable in some places by the useless 'disturbed' peat that always overlaid it. When he shewed us a small part of the trackway it, like the Meare Heath track, lay upon the surface of the old humified *Sphagnum–Calluna* peat and was covered by *Cladium* peat, the 'bad' peat he had referred to. When, in 1944, we were able to excavate and record a section of this, that we called 'Foster's track' or 'Shapwick Heath track', we were able amply to confirm this general stratigraphic conclusion. Furthermore the tree-pollen diagram showed clearly that the trackway occurred at the end of our zone VIIb, the Sub-boreal period, and the opening of zone VIII, the Sub-atlantic period. Although indeed the transition was by no means precise, the pollen zoning lent support to a Late Bronze Age date for the trackway, and hence to the events that induced the stratigraphic change at trackway level. This can only be described, in the light of the evidence, as a substantial flooding or swamping of the surfaces of the raised bogs by calcareous water from the river catchment. To judge from the consistency of the dark lower peat, the bogs must have been remarkably dry when it formed, and its thickness surely indicates that for a very long time, probably a thousand years or more, they had suffered no comparable inundation.

We were naturally anxious to discover the history of the bogs, with their possible intimations for climatology and archaeology throughout the time of formation of the upper peat, but the widespread peat cutting although still revealing frequent fresh evidence, had already destroyed much of the upper peat, so that by the early 1940s there were few places where more than 3 ft (1 m) of upper peat remained, and even some of this was liable to be turf reworked from a previous peat cutting. We

Fig. 35. Profile through the upper peat on Shapwick Heath shewing the growth of ombrogenous *Sphagnum* peat interrupted by two phases of flooding by base-rich water, during which the giant sword sedge (*Cladium mariscus*) was dominant. These flooding episodes were later widely recognised in the region, the earlier being strongly associated with Late Bronze Age trackways and platforms.

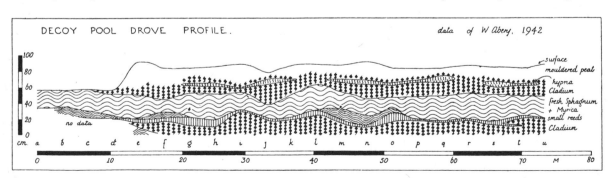

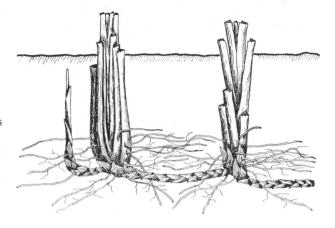

Fig. 36. The giant sword sedge, *Cladium mariscus*, shewing the massive bases of its aerial shoots and substantial rhizomes, both readily recognisable in peat, the latter bright salmon-pink in colour. (Drawing by R. H. Yapp.)

counted ourselves lucky, therefore, to find in 1942 a deep freshly cut face of the upper peat at the edge of Decoy Pool Drove close to the site where we had examined the Shapwick Heath track. This face was most carefully measured and recorded by Miss W. Abery, a recent Cambridge graduate then teaching in Street. Figure 35 gives a synopsis of the striking results she obtained. It was quickly confirmed that the whole of the deposits shewn rest directly upon the surface of the lower humified peat whose upper surface is black and crumbly with twiglets of sweet gale and much detritus, indicative of a flooded bog surface. The long section clearly shews firstly, the prevalence of the *Cladium* peat, easily recognisable by its stout black rhizomes with a core of red vascular strands and by its urn-shaped fruits showing, when divested of the enclosing utricle, three stout cusps at one end. Secondly it is apparent that the upper peat contains two thick and distinct layers of *Cladium* peat, separated by a substantial layer of oligotrophic *Sphagnum–Calluna* peat, largely fresh in character and containing such typical bog plants as cranberry, bog-asphodel, both species of cotton-grass, bog andromeda, and frequently the bog-bean. Locally the lower part of this intermediate peat shewed lenses of more highly humified *Sphagnum–Calluna* peat, and below this, generally as a transition from the sedge peat to the *Sphagnum* peat, was a layer of the small reeds, such as *Phalaris* or *Calamagrostis*, along with some *Molinia*, together constituting just such a community as occurs in the succession from a shallowing *Cladium* fen. We were now constrained to modify our original simple version of the nature of the Boundary Horizon derived purely from ombrotrophic bogs and to accept that in a geographic region such as this, climatic worsening had naturally resulted in flooding episodes during which fen communities extended between and over the raised bogs, altogether altering their qualities as habitats for plants, animals and humans alike.

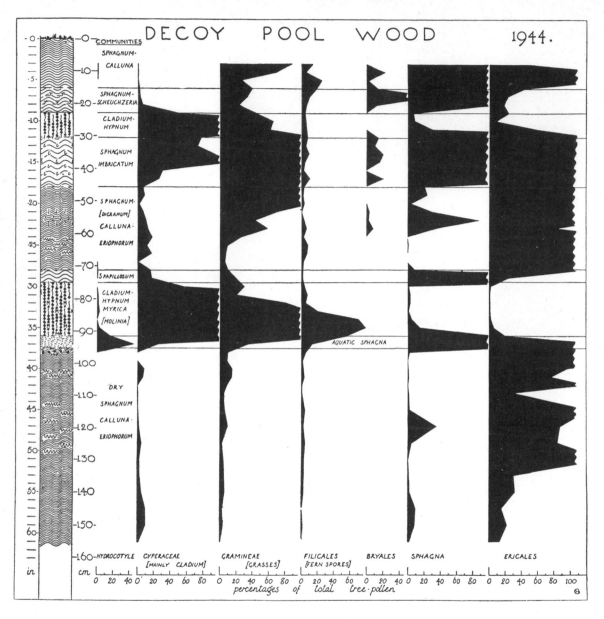

Between the two flooding episodes it is clear that we have to postulate a return to conditions of infrequent or no flooding, during which the growth of the ombrotrophic bogs was resumed.

We now were under compulsion to satisfy ourselves that this sequence of Sub-atlantic events applied widely throughout the Levels and looked for other suitable thick exposures of the upper peat. One such we found on the edge of Decoy Pool Wood where the bogs of the old

Sphagnum–Calluna peat lay at a depth of 95 cm. The peat stratigraphy most clearly illustrated the dual flooding episodes and shewed reversion after the second of them once more to ombrotrophic peat. The sequence is well displayed in the diagrammatic section of Fig. 37. Here the non-tree pollen derived from the local vegetation is also displayed and it will be seen that at each flooding episode the frequency curves for the calcifuge heathers and ling fall to negligible height, whilst the pollen of the sedges (including *Cladium* itself) rises right off the graph, and the grass-pollen curve is similar, suggesting the prevalence of the small reeds along with the sedge. It is interesting to see the sudden increase in pollen of *Hydrocotyle vulgaris*, the tiny marsh pennywort, in the aquatic pool peat that was the first indication of the flooding of the old *Sphagnum–Calluna* peat surface: it is a very typical fen plant, not at all typical of the active raised bog. The combination of our pollen analyses with careful dissection of peat columns in the laboratory gave a good deal of certainty to our general conclusions as to what could be called the 'palaeoecology' of the bogs. It now seemed evident that the flooding by alkaline water on two occasions between the Late Bronze Age and Late Roman time must have affected all the raised bog area of Shapwick and Meare Heaths, and there were indications in the field notes of our first visit to the Roman hoard site that with more experience we should have seen the evidence there also. No doubt I was heavily prejudiced at that early stage against accepting a field identification of *Cladium* in a raised bog!

Recognition of the flooding episodes shed a very welcome light upon a problem that bothered me from the outset. In 1906 there had been found in widening the drain beside the Shapwick to Westhay road, and close to the old Shapwick railway station, a substantial dug-out boat of oak: it had been extracted from the peat and now rested in Glastonbury Museum. So far as one could see from examining the ditch sides the site was part of the Shapwick Heath bog complex, but whatever was a large monoxylous boat (5.2 m in length) doing on the top of the raised bogs? The bog pools and lakelets in such mires offered no answer, but at either

◄

Fig. 37. Detailed analysis of peat stratigraphy and of non-tree pollen on Shapwick Heath, Somerset. Two flooding episodes are recorded by layers of *Cladium* peat, the oldest on the surface of the old *Sphagnum–Calluna–Eriophorum* peat. At each flooding horizon the ericoid pollen almost ceases and there are huge maxima of sedge and grass pollen indicative of giant sedge and reeds.

Fig. 38. Key to the peat stratigraphic symbols generally employed in pollen diagrams and bog profiles.

A | Fresh *Sphagnum* peat | Fresh *Sphagnum imbricatum* peat | *Molinia* peat | Wood peat
| Highly humified *Sphagnum* peat | *Eriophorum vaginatum* | *Molinia–Sphagnum* peat | *Myrica gale Andromeda polifolia*
B | Aquatic *Sphagnum* peat | *Phragmites* peat | Hypnoid moss peat | *Carex–Sphagnum* peat
| *Scheuchzeria palustris* | *Carex* peat | *Juncus* peat | *Cladium* peat
C | Calcareous necron mud | Shelly necron mud | Coarse detritus mud | Drift mud and reworked mud
| Necron mud | Silty necron mud | Fine detritus mud |

Plate 42. Peat section near Westhay, Somerset, shewing clear separation between the lower highly humified *Sphagnum–Calluna* peat and the upper less humified peat. Turves of the latter shew the pronounced banding of the regeneration cycle. Some distance above the boundary between the two major peat layers (the first swamping horizon), cracks and the gaping of cuts made by the peat spades shew the contraction of the sheet of aquatic pool peat formed in the second general swamping episode. (1948)

of the two flooding episodes the tops of the Shapwick–Meare bogs were awash in a sea of flood-water though bordered and even covered with beds of sedge and reed. The whole region must have been navigable to shallow craft, which could of course move along the deeper drainage channels between the bogs, and thus were in open-water contact with the great lake north of the island of Meare, with its Iron Age and Romano-British lake-side villages. When, much later, we were able to obtain a radiocarbon date for the timber of the dug-out boat, its reference to the Iron Age was confirmed.

Widespread as the two flooding episodes were, there were some parts of the Levels too high to be overtopped by the calcareous water. One such region lay to the west of Meare Island, where between Westhay and Catcott Burtle extensive peat cuttings were being made. In the seasons 1944 and 1945 although we inspected very long peat faces in pasture that had apparently not been previously cut for peat, we saw no trace at all of eutrophic plant communities in the upper peat. None the less traces of the two flooding episodes were apparent in the form of layers of aquatic *Sphagnum* peat, the lower lying directly upon the surfaces of the old humified *Sphagnum–Calluna* peat, and the upper upon an ombrotrophic bog peat sometimes shewing very pronounced

Plate *43*. The first cuts in the surface of a former raised bog long used as a meadow. Right across the bog is a very conspicuous white layer of the papery peat formed by the rhizomes of *Scheuchzeria palustris*, a certain indicator of extensive flooding by base-deficient water. (Near Westhay, Somerset, 1963)

lens structure of the regeneration complex. The upper flooding horizon contained the abundant papery rhizomes of *Scheuchzeria palustris* that I had seen growing in the bog pools on the great Swedish raised bog, Komosse. It is so limited to this ecological niche that it infallibly indicates the open pools of acidic bogs. It was apparent here in 1944, 1945 and 1947, at the last date recognisable as quite a thick pale fibrous layer along the whole peat face, and it was from this layer that the keen eyes of Frank Mitchell, then visiting our sites, soon detected the fruiting heads of the plant. We encountered this plant also in the upper swamping surface at Decoy Pool Wood, at 'Sandford's track' near Westhay, and much later at the reopened Shapwick boat site.

The Westhay track site (Sandford's track) was of interest not only because of its intermediate status in the range of flooding, but because it gave us our first opportunity to examine the construction of a substantial prehistoric trackway and at the same time provided its own indication of the archaeological period in which it had been made. In the short piece of track that Roy Clapham and I excavated, it was substantial in construction, about 6 ft (2 m) in width and built around stems of birch timber trimmed of their branches and laid longitudinally in the direction of the track. Brushwood had been laid between and over these main timbers, which were up to 6 in (15 cm) diameter. The brush included ling and bog myrtle, and was pinned down by sharpened vertical stakes, usually about 1 in (2.5 cm) in diameter, though sometimes larger. There was little or no transverse wood but it must have had

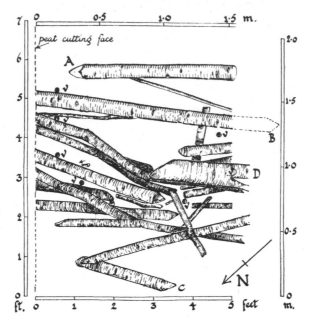

Fig. 39. Westhay track, excavated in 1944. Plan shewing the main longitudinally laid birch timbers and small, more or less vertical stakes pinning the brushwood (now removed) of the track to the peat below. The ends of the timbers A, B and C displayed axe cuts (Fig. 41).

considerable strength, as may be judged from both plan and section (Figs. 39 and 40). The latter shews the track, like the others so far mentioned, sitting upon the surface of the old *Sphagnum–Calluna* peat, covered with a greasy laminated layer of aquatic *Sphagnum* peat, this with many tussocks of cotton-grass, but both sealed in by a thick bed of *Cladium* peat full of the red-cored rhizomes at the top and fibrous rootlets below, whilst its fat unbranched black roots capable of growing into deep muds devoid of oxygen had penetrated through the track and into the old humified peat. One could not have desired clearer evidence of the *Cladium* sedge-fen overwhelming the old bog surface. Above the fen peat there was a variable thickness of *Sphagnum* peat, indicating by its strong lens structure the alternations of the pools and hummocks formed during active regeneration of the returned raised bog. Finally this was succeeded by an aquatic *Sphagnum* peat containing *Scheuchzeria*. Thus we had a site where the first flooding led to eutrophic fen: in a second shallower flooding calcareous waters failed to overtop the bog, which none the less was covered with acidic bog pools in response to 'worsening' conditions.

As we finally took the track apart after measuring and drawing it, Roy and I sought to see how the timbers had been cut by gently washing clean the cut ends of some of the largest. We were all the same surprised to see how clearly the cleaned surfaces displayed the marks of the felling axes. These had quite evidently been thick-bladed with strong lateral

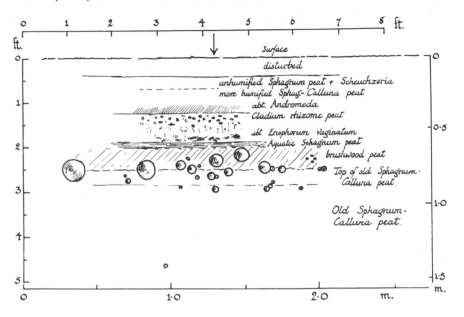

Fig. 40. The Westhay track (*F* in Fig. 34) seen cut across in the peat face (1944). *Cladium* sedge peat, indicative of a major flooding, overlies the timbers of the trackway that has been built on the top of the old humified *Sphagnum–Calluna* peat. A later flooding episode is indicated by a layer of *Scheuchzeria* in the upper peat.

cusps and a strongly curved cutting edge, indeed almost certainly they were the hollow-cast bronze axes of the Late Bronze Age. So clear were the cuts on the well-preserved birch that we could measure the exact width of the chord in many of the blade marks: as this dimension was 4.0 cm in one timber and 5.0 cm in another, at least two such axes were at work on the track. This direct age attribution to the trackway was consonant with the fact that the peat digger had earlier recovered a hollow-socketed, looped spear-head of the Middle Bronze Age from a site only 245 ft (75 m) away from the track, and at a depth recognisable in terms of the careful turbary practice, corresponding with about 9 in (18 cm) below the trackway (Fig. 45).

It was of course in our minds that the abundant mortise holes, often cut in good rectangular section of the massive Meare Heath track, were strongly indicative of metal tools such as those of the Late Bronze Age.

During the next few years we were able to identify and examine several further instances of wooden trackways in the bogs of this limited region. It greatly surprised us, coming fresh to the turbaries with a primary interest in the bog stratigraphy and evolution, that we so readily found and so willingly were offered information by the peat diggers about possible occurrences of trackways. It seemed we had come at a time when the possibility of recovering wooden artefacts from the turbaries had scarcely dawned upon local archaeologists. I have to make instant qualification of this comment since, no sooner were we put in touch with H. S. L. Dewar, a retired tea-planter living on the Polden

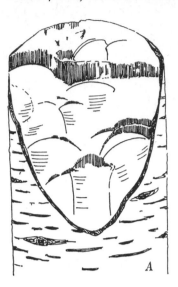
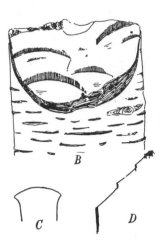

Fig. 41. The ends of two of the larger timbers of the Westhay track, when cleared of peat, shewed marks of thick, curved-bladed axes such as those used in the Late Bronze Age (C). Measurement shewed that different individual axes were employed for cutting stems *A* and *B*. *D* shews cuts in profile.

ridge, than he became an entirely enthusiastic colleague, cooperator and independent investigator, on whom we greatly relied for information of all kinds, especially of early notice of new discoveries.

We have already described the Shapwick Heath track as a slight structure, with no evident core of main timber. It seemed to have been made of brushwood, containing ling and bog myrtle, reinforced by sparse horizontal stems of hazel, all pinned down by similar stems, only about 2.0 cm across and 10.0 to 20.0 cm long but sharpened and driven into the underlying black peat at various angles. We found no residual peat that would allow us any detailed excavation, but the stratigraphic position of the track was quite clear and two of the cut stakes proved on microscopic examination to be of beech (*Fagus sylvatica*), which accordingly had to have been already growing here as early as the Late Bronze Age, though at that time doubtfully regarded as a native British tree. Other woods identified were those of hazel, ash, hedge-maple and guelder-rose.

Another trackway of relatively slight construction was that which we called Toll Gate House from its proximity to a building at the north end of the causeway road across the River Brue from Westhay village. Here also we were unable to excavate the track properly, although we obtained a careful stratigraphic record and a good pollen-analytic series from where, in 1947, we found it exposed in a recent peat face. It lay so precisely upon the projected line of a trackway we had already excavated half a mile (800 m) to the north (Blakeway Farm track), that we assumed it to be the same one and neglected further examination at the Toll Gate House site. None the less the track was shown to lie at the surface of a highly humified *Calluna–Sphagnum* peat and to be immediately covered by 30 cm of aquatic *Sphagnum* peat, and as the tree-pollen zonation proved the track to lie between zones VIIb and VIII we concluded that here again we had the consequences of the first flooding episode. In the peat face the track was seen to consist of a thin horizontal row of straight hazel stems of about 1 to 2 in (2.5 to 5.0 cm) diameter, making a pathway about 2 ft 8 in (80 cm) width. When encountered accidentally in the following year the track was, however, of much more robust construction, possibly in response to variation in the bog surface at some locally wetter spot. Our assumption of a probable Late Bronze Age origin was confirmed by the clean-cut ends of the hazel stems two of which bore the marks of small curved-bladed axes.

It was also in 1947 that, in company with two former research students, J. N. Jennings and P. A. Tallentire, another trackway was detected in deep peat cuttings east of Decoy Pool Wood. It disappeared into a steep face of uncut peat about 6.7 ft (205 cm) below the same apparently untouched bog surface as that from which the Late Roman hoards had been buried. We seized the opportunity to secure pollen samples and these were extended by the chamber borer down to the surface of the underlying estuarine blue-grey clay. The peat diggers were in no doubt that the large timbers we now saw in the section were part of a trackway, whose direction was said by them to have been indicated by pairs of heavy piles mortised at their upper ends, and occurring at intervals in a straight line over a considerable distance. All such timber had unfortunately now gone. I recall being left through the hot day to work on the peat section, and how I sought to scramble up the bank at lunch-time, narrowly avoiding putting my hand upon an adder sunning himself on the stump of a willow that looked to be a convenient hand-hold. When this had happened two or three times it was borne in upon me that the snake had prior rights, and failing other identification this trackway became 'Viper's track'. We were fortunate that we took trouble over this very favourable exposure, for although it was impossible to excavate here, two years later Dewar encountered the trackway some 820 ft (250 m) south, apparently upon the rand of the raised bog. It was evident as cut timber in the turves and peat faces of a number of parallel cuttings, and with the help of members of the Glastonbury Antiquarian Society, an elegant excavation of the intact trackway was made of some 12 ft (3.65 m) of its length. A very careful scale plan of the main features of the trackway was made by Mr G. H.

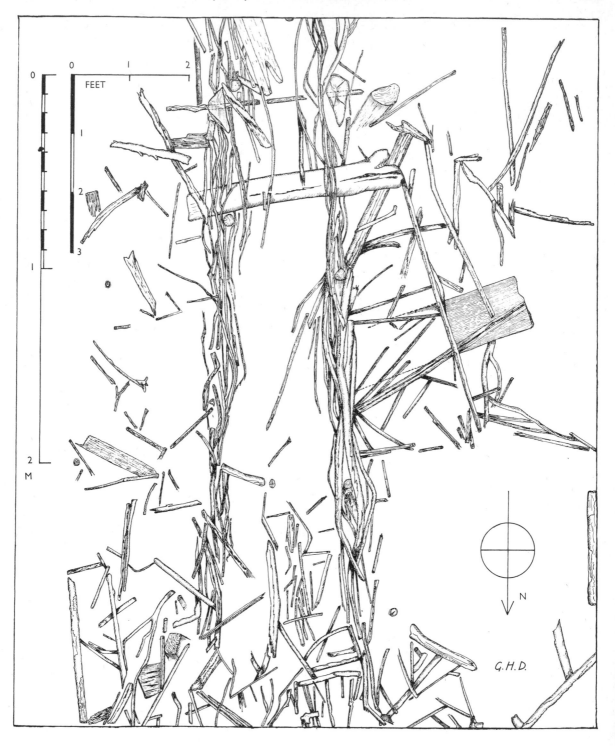

Davis and is reproduced in Fig. 42. The track was quite a substantial structure, ranging in overall width from 6 ft (2 m) to nearly twice this where it seemed to have splayed out laterally in use or to have been repaired. It consisted of a core of longitudinal wooden stems, 1 to 4.5 cm in diameter to a total (surviving) thickness of about 18 in (46 cm). In some places these cut branches were supplemented or replaced by much more substantial timbers including even squared oak or small tree trunks. Thick timber of this kind was found placed or inserted in various directions, oblique, longitudinal or transverse. The main mass of the track was pinned through to the underlying black peat by short sharpened stakes, as we had observed in the previously excavated trackways. Two features, however, specially characterised the Viper's track. The first was the presence of double pairs of stout piles at intervals, recorded as between 10 ft (3.05 m) and 17 ft (5.2 m), along the trackway: these were no doubt what the turbary workers had reported to us. The piles were of heavy timber now seen projecting well above the brushwood core, each perforated by a single large mortise hole that in some instances could be seen to carry a transverse bearer or tie-rod beneath the longitudinal stems. We found it impossible to judge whether the doubling of each pile had been a feature of primary construction or a secondary one intended to offset the splaying out of the track in heavy use or flooding. The further constructional feature new to us was the presence along each side of the track, between the piles, of stringers made up of twisted bundles of slender rods interlaced about upright pegs or stakes about 1½ in (4 cm) in diameter. The usable track between the lateral stringers was only 2 ft (50 cm), despite the much greater width of the whole structure. This trackway was visible in peat cuttings many times up to 1959, and was demonstrated to an excursion of the British Association at its Bristol meeting in 1955. On all these occasions it was extremely easy to make out the local peat stratigraphy and it was beyond doubt that the trackway was built on the surface of the black, very highly humified *Sphagnum–Calluna* peat, and that it was covered by blackish-brown *Cladium* peat with stout rhizomes and abundant black roots descending through the detritus peat of the flooded surface. One could often make out above this first flooding horizon a return to ombrotrophic bog and the presence of another *Cladium* peat of the second flooding episode, but this sequence was exhibited *par excellence* in the records secured at the initial 1947 exposure, where furthermore the detailed pollen analyses lent every support to the stratigraphic evidence and confirmed the trackway's provenance as the Sub-boreal/Sub-atlantic transition. On several of the stouter timbers there were abundant and clear tool marks consistent with the use of an

◀

Fig. 42. Plan of the excavation of Viper's track, 1949. Near the top of the drawing note the substantial paired piles with a transverse bearer between. The track is limited by lateral stringers of plaited branches, and the whole structure has clearly suffered a fair degree of damage before it was preserved.

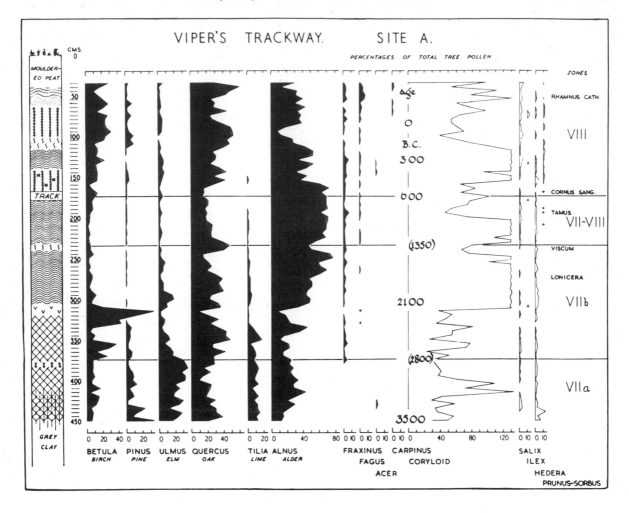

VIPER'S TRACKWAY.   SITE A.

PERCENTAGES OF TOTAL TREE POLLEN

axe of Late Bronze Age type with strongly curved blade about 1.5 in (3.8 cm) from one cusp to the other. Viper's track had thus yielded an extremely full and consistent variety of evidence, though leaving still unanswered questions of the extent of damage and repair.

We had not yet seen the end of the revelations of this very productive decade. Whilst Viper's track was being excavated in 1949, H. S. L. Dewar detected evidence of yet another trackway exposed in the same peat face. This, the Nidons track, ran north at an angle of about 35 degrees east of Viper's track, with which it seemed by extrapolation to have converged in the flat meadow where the lagg of the raised bog must formerly have been. Dewar excavated the trackway in 1953 near to its original exposure, and opened another view of it in 1955 for the British Association; later, in 1959, Miss J. Turner and I sketched and photographed the track in yet another peat face where we extracted one of the stout vertical piles showing typical blade marks of the small Late Bronze Age axes. Nidons track very much resembled Viper's track in construction, although lacking the lateral stringers of twisted rods. In both, the piles consisted of wide cleft boards of wood shaped to a narrow triangular apex, mortised and used in pairs driven into the peat obliquely to converge beneath the trackway. The trackway seemed to have been even more damaged than Viper's track had been and repairs had been extensive: possibly these structures were specially susceptible on the curved rand of the bog where running water at the time of the swamping must often have had maximum destructive effect, directly and indirectly. One should add that in all exposures of Nidons track it was quite clear that it had the same stratigraphic provenance as Viper's track, i.e. on the surface of the highly humified black lower peat.

It was remarkable that we recovered yet other evidence of wooden artefacts from this stratigraphic horizon. Midway between the two trackways, and where they were about 400 ft (120 m) apart, J. N.

◀

Fig. 43. Viper's track, Shapwick Heath. The peat stratigraphy shews the trackway at the top of the old *Sphagnum–Calluna* peat and just below the *Cladium* sedge peat of the first flooding episode. A second flooding episode is indicated at a higher level. The large shifts in tree-pollen frequencies allow zoning of the diagram with ombrogenic peat developing in zone VIIb and the trackway being built at the commencement of zone VIII. The radiocarbon ages are a later interpolation, but confirm the zoning.

Plate 44. Nidons track, Shapwick Heath: one of the vertical oak piles shaped to a point by Late Bronze Age axes that have left characteristic markings.

Scales: centimetres and inches.

Plate 45. Viper's platform, seen transected by peat diggings on Shapwick Heath, 1947. This structure was subsequently excavated. It was one of several, all of which lay at the surface of the old *Sphagnum–Calluna–Eriophorum* peat and were accordingly presumed to be of the Late Bronze Age.

Jennings and I encountered in the peat faces what we took at first to be sections of yet another trackway. These, however, did not link into the expected straight line and when excavated they turned out to be isolated platforms of stout alder logs about 4 in (10 cm) in diameter laid separately parallel with one another above a mattress of small bundles of birch placed alternately across one another. This mattress was secured not only by a multitude of sharpened small stakes driven through it obliquely in all directions, but by horizontal stems driven laterally through, and a marginal row of stakes driven obliquely downwards and inwards. At one end only, the alder timbers had been so deeply notched that we supposed they must have supported some structure such as a wall or screen. No associated artefacts were found and it seemed likely that these structures, of which we saw evidence of three or four, were look-out platforms from which both trackways might be kept in view by an observer looking over the top of the giant sedge that must then have surrounded them. The position at the top of the bog margin would have been very advantageous for this. Alternatively they may have had use in wild-fowling or related activity. Once again we saw evidence of tool marks of the Late Bronze Age axe type.

Although it was evident that it would be most desirable to know how these trackways behaved as they left the curved bog margins and crossed the wet lagg lying next to the hill-slope, and especially to find

Plate *46*. Viper's platform. About one-third of the original structure remains from previous peat cutting. The platform was made of large horizontal timbers placed parallel upon a substantial brushwood mattress and deeply notched at one end as if to carry a transverse structure such as a screen. The whole was securely pinned to the underlying highly humified lower peat. The horizontal staff is 1.5 m in length.

the point of convergence of Viper's track and Nidons track, the pasture in this position was so wet that excavation was impracticable. Dewar did indeed succeed later on in recovering remains of a wooden structure in this locality, 'Tully's track', but the large aggregate of miscellaneous timber found upon excavation disclosed little in the way of organised structure.

In the relatively short period between 1941 and 1950 the turbaries of the Shapwick region had now provided us with an almost embarrassing wealth of evidence, derived in the main from the upper unhumified peat at that time within easy reach of the turf cutter's spade. Leaving aside for the moment the less frequent and still largely obscure discoveries we had begun to make within the older and deeper peat, it seems most rewarding to devote our attention now to the organisation of the data so far gathered and to consideration of its many implications for all the various fields of enquiry that are involved. The next chapter is devoted to this helpful exercise.

9

Trackways in context

Principles of trackway construction

In 1951, when we turned to take stock of what had been achieved up to that time in our study of the wooden trackways of the Levels, there had nowhere been any sustained study of the principles of construction, location or purposes of such structures. We were now in a position, however, to make effective comment on these issues at least with regard to the Late Bronze Age, to which it seemed all the trackways so far described must belong.

We enjoyed of course the great advantage of familiarity with the topography and surface ecology of the living surfaces of active raised bogs such as these over which the trackways were laid down. Perhaps it was this experience that allowed our first generalisation: the trackways were not bridgeways, supported upon heavy piles to keep elevated viaducts clear of bog or swamp below. Here they unquestionably had the character of wooden and brushwood mattresses pinned down to the bog surface by short sharpened wooden stakes. Their purpose was primarily that of strengthening the soft bog surface, extremely susceptible as it is recognised to be by anyone who stands about on it for long, even if only to take a peat boring or make a vegetation map. These longitudinal mattresses found their precise parallel in the enthusiastic adoption by the Somerset peat industry after the 1939–45 war, of the elongate perforated metal sheets that had been much employed to create temporary aircraft landing strips over surfaces too soft otherwise for such use. No doubt I was myself specially aware of the necessity for surface strengthening from the fact that my own feet are small for a biggish man, and would often break through the top vegetation where my companions had trodden safely, a result made more likely by my carrying around heavy spades, peat augers, rucksacks and such gear.

The mattress principle once enunciated, it was evident in the manner of both construction and repair of the trackways. These clearly varied from extremely light, as in the Shapwick Heath and Toll Gate House tracks, through to substantial affairs like Viper's, Nidons and the Westhay track, and finally to the still more massive structure of the Meare Heath track. In all of these except the last, a linear construction was predominant in the timber core. Whether we are concerned with small or large diameter rods, this construction is peculiarly unfitted to be trodden by the narrow human foot that would so easily be trapped between the parallel surfaces; nor would it be any better for a sled or wheeled vehicle for that matter. On the other hand, longitudinal timbers effectively bridge the numerous pools and drainage channels that we know to be so much a feature of the living raised bog surface: a transverse arrangement to overpass the same pools would require to be built wider, with a far greater expenditure of material and labour. There is abundant evidence that the central core of longitudinal timbers was underlaid and packed by smaller brushwood. It seems highly likely that this brush, perhaps together with peat sods also, made a compact trackway surface, kinder to the tread, although trace of this is generally gone by wear and oxidation.

We had still to rely for description of the Meare Heath trackway on the account Bulleid had supplied in 1933: our own contribution had been simply to relate it to the stratigraphy and pollen zonation, and to extend the line of its known occurrence as far as the southern flank of Meare Heath. We were not able to redetermine the trackway construction, though the little we saw in marginal exposures did not contradict what Bulleid had written. The massive construction of transverse flat timbers, and the presence (as he wrote) of large straight longitudinal stringers held by upright pegs at the two sides of the track, over the sleepers, led naturally to the conjecture that this trackway at least, might have been meant for sleds or wheeled traffic.* There was, however, no direct evidence of this and clearly the majority if not all of the trackways were for foot-passengers. They were of considerable length, passing in straight lines from one low island or hillside to another, thereby implying considerable manual effort, the more so because the vast majority of the trees employed grew no nearer than the bog margins, and a great many were represented by substantial timbers of oak, maple and ash that favoured the mineral upland soils and were both tough to work and heavy to carry.

We are helped by the knowledge that the surfaces of intact raised bogs which are in the heath-clad state of *Stillstand* or erosion complexes (as

* Not until 1977 was this conjecture finally disposed of by the renewed work of Coles and Orme: they have proved that some tracks at least were surfaced by longitudinal pairs of flat planks, and that this was so in the Meare Heath track, where, much displaced, they had been taken by Bulleid to be stringers.

these particular raised bogs were in the Middle to Late Bronze Age) can be readily traversed on foot, so that they would have offered little obstacle to communication across the Levels between the Poldens, the Wedmore Ridge and the various intervening low islands. It is easy to understand that when the general flooding with calcareous water extended almost completely over the existing complex of raised bogs, such routes were heavily jeopardised. We may readily perceive how, as water-logging progressed with the deterioration of climate, the trackways were built, probably along existing routes across the Levels, so as to avoid the extremely circuitous alternative upland routes. Although there was time for some of the trackways to wear and break up, the majority were entombed and preserved by the rising water and the new sedge peat that it induced. There are no corresponding trackways in the peat for some depth below nor above this one critical Late Bronze Age level: thus in the first great flooding episode we find at once the reason for the building, for preservation and for contemporaneity of the trackways. With this goes, I think, the inference that there was no inherited or imported long tradition of how wooden trackways had always been built in the human society now faced with the deluge, but there was scope for various means of construction to be invented and tried out according to local needs and local materials.

Although it is true that the trackways were buried sufficiently soon for them to be well preserved we often noted features suggestive of repair or strengthening. This was especially apparent in the two tracks (Viper's and Nidons) where they descended the sloping bog margin. In Viper's track especially, whereas the ultimate path between the marginal stringers was only about 2 ft (50 cm) wide, in places the width of the track, augmented by the dumping of miscellaneous material of very various kind, had become as much as 11 ft (3.3 m) or so. It is also most likely that the doubling of each member of the pairs of piles had resulted from driving down of the second members to offset the initial outward splaying of the first pair. The rand was naturally susceptible to destruction because of its slope and because it experienced more water-flow whenever the enlarged lagg-stream was running most strongly. Valuable as no doubt the trackway system was, it could not sustain the overwhelming rise in water-levels and all were abandoned, so that, in a broad sense at least, all the Late Bronze Age trackways were contemporaneous. It was interesting to reflect that were a general climatic worsening to be, as we supposed, the ultimate cause involved, then we might expect to find other Late Bronze Age trackways in other parts of Britain, and we could reflect that the Fenland Research Committee had already confirmed the existence of a massive trackway of this age at

Barway in the Cambridgeshire Fens. This incidentally crossed a substantial river channel and was supported by extremely massive piles. We found no evidence at this time or later, in the Somerset raised bogs, to suggest that the wooden trackways were in any way associated with harvesting of any produce or crop from the bog surface, as was apparently the case with tracks built upon the north-west German bogs to facilitate removel of pine timber from them.

Correlation of events

In pursuing such interdisciplinary studies as I have been describing, one constantly looks for the results of the various lines of enquiry to assist and complement one another. Above all, however, the conclusions from the different sources must agree: it must be possible to arrange the facts and ideas in one general scheme whose validity is guaranteed by its all-round applicability. By 1951 such a unifying pattern was indeed becoming clear to us.

Bog stratigraphy had been shewn to be very consistent. Everywhere in the Somerset bogs there was exhibited the abrupt division of the raised bog peat into a lower dark highly humified *Sphagnum–Calluna–Eriophorum* peat and a pale fibrous and unhumified upper peat, the two separated by aquatic peat, formed as a consequence of a major flooding episode. The upper peat, generally acidic *Sphagnum* peat, was itself split into two by the aquatic peats of a second flooding episode apparently less severe than the earlier one. This is the sequence represented

Fig. 44. Stratigraphy at the site near Shapwick Station, where the dug-out boat was found in 1906. It will be noted that here, as elsewhere on Shapwick Heath, there are two flooding horizons indicated by *Cladium* sedge peat. The lower, in which the boat was probably found, sits directly upon the surface of the humified lower peat: this suggests a pre-Roman Iron Age date for the boat, as radiocarbon dating confirms.

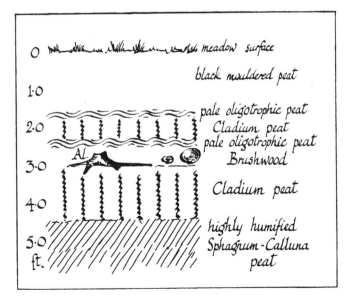

throughout the area with the qualification that the aquatic peat of either flooding consisted primarily of sedge peats in the lower levels and nearer the Poldens, whilst on higher-standing bogs it was the peat typically forming in large oligotrophic pools.

We were inclined to see this sequence as synchronous throughout the Levels partly because of its very great lateral extent in peat exposures and partly because we regarded widespread climatic deterioration as its most likely cause. Despite much search we never found any evidence for an alternative possibility, viz. that either episode was a consequence of a rise in sea-level having induced backing-up of fresh water as was the case in the East Anglian Fenland.

Pollen analysis gave general support to the framework of our correlation. The analysis of tree-pollen represents the history of the plants of the uplands, rather than of lowland bogs and fens, and therefore of communities where the dominant plants, the trees, might be supposed to be most directly responsive to climatic control and thus be the indicators of major changes of climate over the whole country or at least the entire region. Since 1940 we had been employing a provisional pollen zonation for England and Wales in which a consistent pattern of change was used to mark the boundary between zone VIIb, thought to correspond with the Sub-boreal climatic period, and zone VIII, thought to coincide with the Sub-atlantic climatic period. The indices of change at this boundary were the termination of a continuous record for pollen of linden or lime (*Tilia*), a sustained increase in the frequency of birch (*Betula*), the appearance and sustained presence, especially in eastern England, of beech (*Fagus*) and, to a smaller degree, of hornbeam (*Carpinus*). Attempts to define these changes more closely by making a

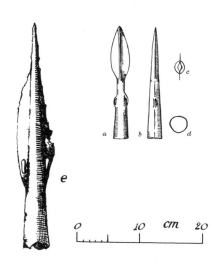

Fig. 45. Two Middle Bronze Age spear-heads from known stratigraphic context in the Somerset Levels. *a–d* shews that found near the Westhay track and very probably from a similar horizon. *e* is the Coppice Gate spear: the extensive one-sided corrosion suggests that it may have remained half-embedded in the bog surface for some time before being covered by peat growth (see Plate 47).

Plate 47. Hollow-cast spear-head in the position of its discovery near Coppice Gate, Shapwick Heath, 1949. The spear is within the upper part of the old *Sphagnum–Calluna* peat seen to be penetrated by descending rootlets of *Cladium* from the overlying *Cladium* sedge peat, formed in the first flooding episode of the region. The peat face cleaned in part for examination and sampling: pollen analyses confirmed the indications of the peat stratigraphy and also shewed considerable upland agricultural activity about the time of the flooding episode.

transition zone VII–VIII had been tried both in the Fenland and in Somerset, but the result was not altogether convincing. What, none the less, was apparent in surveying our Somerset results in 1951 was that the broad pollen zone boundary VIIb/VIII was recognisable in every long series of pollen samples, and that it coincided with the boundary between the lower and upper peat. It was well exhibited in the tree-pollen diagrams for Shapwick Heath (1936), Decoy Pool Drove (1942), Westhay track (1944), Decoy Pool Wood (1944), Toll Gate House track (1947), Viper's track (1947) and, less completely, Meare Heath track (1942). These togcther made a formidable concordance of evidence, by which the major events of the bog stratigraphy were tied to the shift of forest composition on surrounding uplands, and through which the sequence could be extended via the pollen zonation to the rest of Britain and even as far as the lowlands of western Europe.

The most dramatic evidence contributing to the correlation framework for the upper peat by 1951 was certainly that provided by archaeological finds. To begin with, almost every wooden trackway unearthed had displayed tool marks that could have been made only by the very thick-bladed axes with strongly curved edges and pronounced cusps, that were made in profusion during the Late Bronze Age. Such axes were hollow-cast with a central cavity, square in section, to take the sharply angled haft: it was inevitable that the blade itself at the solid

Plate *48*. Four biconical amber beads found near Lytheat's Drove, Westhay Moor. They were about 23 cm down from the top of the old *Sphagnum–Calluna* peat, that is here terminated by aquatic *Sphagnum* peat with *Scheuchzeria*, the evidence of the first flooding episode, during which the non-tree pollen was indicative of substantial agricultural activity. These features accord with the archaeological reference of the beads to the Middle or Late Bronze Age.

apex of the casting should still be very substantial. The abundance of mortise holes in the trackway timber was no doubt attributable to the use of such tools, and in some instances metal gouges had also been deduced. In two separate sites, one near the Westhay track and one on Shapwick Heath, a Middle Bronze Age cast spear-head had been found. The former was a short distance below the upper surface of the old black peat, as we have already described. The second, which occurred on the south flank of Meare Heath at a site we called Coppice Gate, had characteristic lateral loops within the base of the blade itself. We reached the site in time to see the precise mould of its shape in the peat face and it was clearly within the much oxidised and disturbed surface of the lower peat. The blade had been heavily eroded away on one side, as if it had remained exposed on the peat surface for a considerable time, possibly whilst some of the peat had decayed away.

Finally among the chance discoveries was a cluster of four biconical amber beads found in 1951 in peat digging beside Lytheat's Drove, Westhay Moor, on a turbary being worked for Mr E. J. Godwin, who made us very free to examine the site and was so hospitable with refreshment that collecting samples from above the water-logged peat trenches had memorable hazards. These were survived, however, and the bead horizon was clearly identified as between 6 and 12 in (15 and 30 cm) below the contact of the first flooding episode. The beads, a characteristic artefact of the Late Bronze Age, thus again fell at the expected horizon, an attribution later confirmed by the tree-pollen samples taken at the site.

About the time of the last discovery we had report that a bronze fibula, or dress pin with a catch plate, had been found on Shapwick Heath, about 570 ft (174 m) south-east from where the Roman hoards had been recovered. From the finder's description of the depth of the brooch, together with careful examination of the local peat, it was apparent that the fibula had lain in peat of the *second* of the flooding episodes, thus providing our only direct archaeological correlation with that event. Authorities referred the fibula to the Iron Age, probably around the birth of Christ, thus comfortably earlier than the Late Roman discoveries apparently dug into the peat from near the present bog

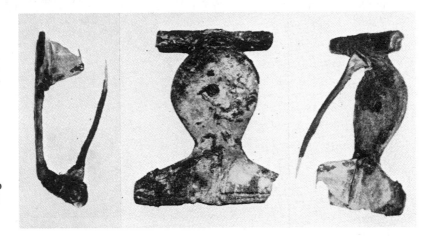

Plate *49*. Iron Age bronze fibula found on Shapwick Heath shortly before 1949. The brooch came from *Cladium* sedge peat attributable to the later of the two main flooding episodes, which corresponds with the attribution, on archaeological grounds, to an Iron Age 3 culture, and possibly the first century A.D.

surface. It is a pity that we were unable to visit the site on Meare Heath where the Iron Age (La Tène) scabbard had been found, until so long after the event that its original level in the peat was obscure and sampling difficult. All the same it seems likely that the find lay in the base of pollen zone VIII and after the onset of the first flooding episode.

It could fairly be said that if internal consistency is the guarantee of the rightness of our time correlation scheme, then by 1951 we had achieved a substantially valid system, in which the evidence of archaeology, bog stratigraphy, pollen analysis, and, behind them all, climatic change smoothly interlocks (Fig. 46).

There were still, of course, outstanding issues. It was disquieting that there was no trace of peat formation on the raised bogs after the Late Roman time. The smooth dome of Shapwick Heath when we first saw it in 1935/6 resembled that of uncut raised bog, nor was there the smallest evidence of peat cuttings down to that surface, though here and there comparatively recent clay drainage pipes had been inserted for agricultural attempts, since abandoned. It seemed that peat formation might well have ceased after the fourth century A.D. If so the cause can scarcely have been climatic since raised bogs elsewhere continued their growth. One lacks evidence of what would seem the most likely alternative cause, namely that it was a consequence of man undertaking drainage of the region.

Advent of radiocarbon dating

The type of chronological framework or reference pattern that we have described so far has the evident advantage that it summarises all the major environmental influences that operate in a given area and shews

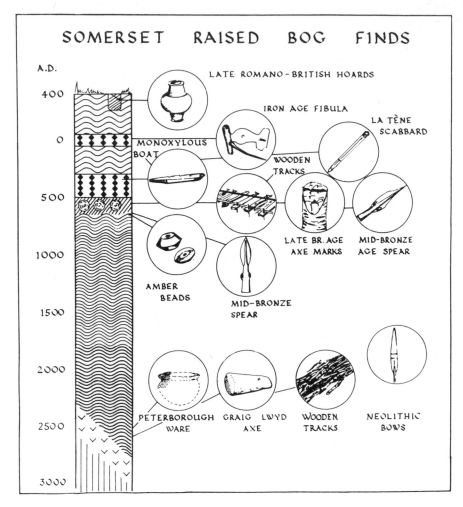

Fig. 46. Schema to shew how the various archaeological discoveries of the Somerset Levels occur in relation to the stratigraphic sequence in the raised bogs of the region. The uppermost series is primarily dealt with in Chapters 8 and 9, the lower (Neolithic) series, in Chapter 11.

how they interact. Equally important, however, is the fact that these influences all ultimately lie under the control of one major external factor, that of changing climate, and because of this the reference pattern can be used as a broad time-scale. The ages established by its use were of course merely relative, although from time to time a link with absolute chronology might be obtained by distant correlation with the archaeology of the Egyptian kingdoms, themselves tied to astronomic events, or to such geological phenomena as the annual laminations deposited in Scandinavian glacier lakes.

The lack of an absolute time-scale for the last 20 000 years or so has, however, been met since 1951, by the development of the technique of radiocarbon dating at the hands of W. F. Libby, working at the time of his original discovery in the Institute of Nuclear Studies in Chicago. This method is now of such extreme importance to the pursuit of all those studies of the foreground of human history that we may reasonably consider the principles on which it depends and how it works.

The cosmic radiation from outer space, meeting the earth's atmosphere generates active secondary radiation in the form of neutrons which form a shell densest at about 40 000 ft (12 000 m) altitude. When the energetic neutrons collide with nitrogen atoms (atomic weight = 14) at this level the balance of seven protons and seven neutrons changes by the ejection of a proton to give a new atom with six protons and eight neutrons. This also has the atomic weight of 14, but, deriving its chemical identity from the protons alone, it is an atom of carbon although the commoner, stable form of carbon has an atomic weight of only 12. The radioactive carbon atom, described as ^{14}C, finally decays by the emission of a beta particle to give again a nitrogen atom. This process of atomic decay is quite random, unaffected by temperature or other environmental conditions, and its average rate is such that a population of radiocarbon atoms loses half its activity in a period estimated at 5730 ± 40 years: a constant known as the half-life.

The active ^{14}C atoms are no sooner created in the atmosphere than they begin to be oxidised to form carbon dioxide, which, like the chemically identical carbon dioxide derived from ^{12}C, mixes with the atmosphere of the earth and is equally subjected to photosynthesis by plants, equally the radiocarbon from it is built into the structure and reserves of plants, and equally is eaten by the animals that feed directly or indirectly upon the plants. Thus the radioactive carbon comes to reside throughout the whole biosphere. Since the decay rate is invariable and, broadly at least, cosmic radiation remains constant, the total amount of radiocarbon in the world also must be constant. The bulk of it is in the carbonates and bicarbonates of the great oceans of the world, held in loose equilibrium with the atmospheric carbon dioxide, and as the ocean waters mix rather quickly around the globe, we find that the atmospheric concentration of radiocarbon remains at a remarkably steady although very low proportion (1 in 10^{12}) of the ^{12}C. Thus all organic material newly synthesised by the growing plant begins with the same constant concentration of radiocarbon in it. But once fixed in the big organic molecules of the plant (or animal) body, the radiocarbon does not exchange with the carbon of the environment but decays *in situ*. As it decays its activity halves for each successive period of 5730 years, and on this principle of exponential decay, carbon dating depends.

The technique of measurement is far from simple, although vastly improved since the first determinations. The first difficulty lies in the

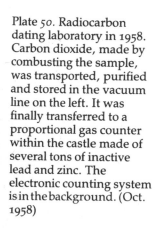

Plate 50. Radiocarbon dating laboratory in 1958. Carbon dioxide, made by combusting the sample, was transported, purified and stored in the vacuum line on the left. It was finally transferred to a proportional gas counter within the castle made of several tons of inactive lead and zinc. The electronic counting system is in the background. (Oct. 1958)

very low concentration of radiocarbon to be assayed, such that 1 gram of carbon *begins* with a disintegration rate of only 15 per minute and the age itself depends on the proportion of this that is residual in the measured sample. The problem is vastly intensified by the fact that the ordinary background radiation from highly penetrating cosmic rays, building materials, rocks and laboratory structures is far higher than the changes we need to measure. One can shield the counting chamber by tons of lead, zinc or steel, but these too are liable to be radioactive; one can work in a laboratory below ground with its attendant disadvantages; and one can seek methods where the counting vessel is small and intercepts as little background radiation as possible. Libby employed the elegant device of surrounding his counter with an 'anticoincidence screen', a cylinder of accessory geiger counters through which all outside (cosmic) radiation had necessarily to pass before reaching the central (^{14}C) counter. By an electronic device any pulse recorded in both screen and central counter was cancelled out, so at a blow the bulk of external radiation was extinguished.

There have been many variants of the counting processes; Libby used elemental carbon, many laboratories employed proportional gas counters with the carbon sample converted to methane, acetylene or carbon dioxide, generally under pressure, and more recently it has become usual to convert the sample carbon into liquid benzene which, in the presence of a suitable 'scintillator' compound, signals each atomic

Plate *51*. Proportional gas counter as used by E. H. Willis in the Cambridge radiocarbon dating laboratory in 1957. It shews the cylinder of overlapping Geiger counters surrounding a Faraday cage which contains the proportional counter itself.

disintegration by creating a point of light that can be registered to constitute the activity count.* By whatever means identified and counted, the result appears as a total of disintegrations per unit time, per gram of carbon, and from this figure, less the comparable rate for dead (inactive) carbon in the same apparatus, the age is easily derived. This age is always given with the laboratory index number and an error term, thus for example: Q-134 Burnham-on-Sea 6262 ± 130 b.c.† It is important to recognise that the error term merely reflects the natural error of counting a population of atoms which disintegrate randomly, so that each count necessarily differs from the next. The error term (1 standard deviation) represents the statistical chance of 2:1 that repeated counts would fall inside the limits shown: twice the magnitude would represent a 19:1 chance. The error term is in no way to be taken as the outside limits of the determined age, nor for that matter is it by any means the only error to which the measurements are subject, whether instrumental or associated with the origin, purity and associative validity of the original samples.

The task of identifying and eliminating all such errors is a fascinating preoccupation of carbon dating scientists, but need not concern us now.

---

* Still more recently there has been demonstrated a method by which the cyclotron is employed directly to measure the proportion of ^{14}C in samples of organic origin.

† Q is the index for the Cambridge radiocarbon dating laboratory: the lower-case 'b.c.' signifies that the result is in uncalibrated radiocarbon years, as are all radiocarbon ages cited in this book.

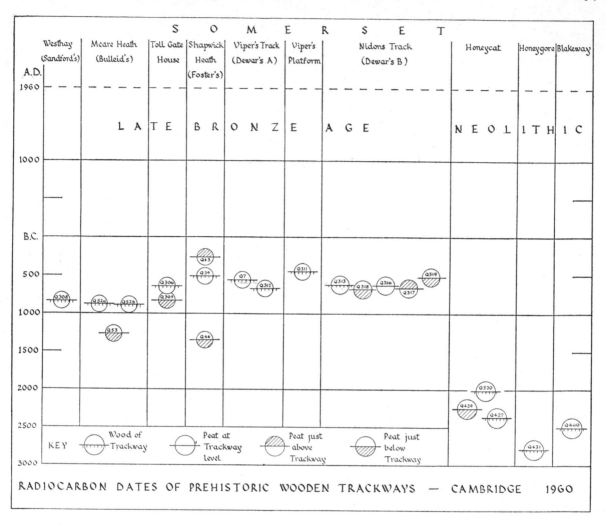

RADIOCARBON DATES OF PREHISTORIC WOODEN TRACKWAYS — CAMBRIDGE   1960

Suffice it that progressively from 1951 the number of carbon dating laboratories has steadily increased and that the annual volume of recorded dates is such that in areas like prehistoric archaeology almost half the contributions to the learned journals are furnished with radiocarbon dates, and commonly rely heavily upon them.

It was by no means always so! When the Cambridge carbon dating laboratory had been built and brought into effective use by E. H. Willis, one of our earliest objectives was to determine the date of the beginning of the Neolithic culture in Britain, where it was particularly characterised by a selective form of disforestation and early agriculture. The generally accepted view then favoured a date close to 1800 B.C., and I recall the discomposure of our leading archaeologists when we offered the radiocarbon dates close to 3000 B.C., thereby at once more than doubling the length of the British Neolithic. Despite their troubled countenances we held firm to the results, which have since been repeated scores and scores of times. Similarly dramatic consequences have attended the application of carbon dating in many other fields, notably to begin with, in that of North American archaeology where the artefacts of early man had been previously impossible to place in any secure time relationship. When Libby first sought to test the feasibility of radiocarbon dating he sought a selection from various parts of the world of organic samples already dated by such independent techniques as comparative archaeology, Quaternary geology and pollen analysis. Among them were two peat samples we had collected from Shapwick Heath: the one (C-347) was peat from the upper oligotrophic layer at Decoy Pool Wood, and the age obtained was 3310 ± 200 B.P. (before present: i.e. about 1360 b.c.); the other (C-343) was taken at the site of our Nidons track excavation from the base of the old highly humified *Sphagnum-Calluna* peat, early in zone VIIa. This latter had a radiocarbon date of 6044 ± 380 B.P.

These results, despite the large errors seemed to us in 1951 to be broadly in accordance with expectation, and by 1959 our own Cambridge laboratory had measured a series of carefully selected samples sufficing to date the upper peat of the Somerset raised bogs and its content of modern trackways. When the wood of the several trackways or the immediately surrounding peat was dated the ages all fell between 500 and 900 b.c., which is certainly Late Bronze Age in southern Britain. Although at Nidons track, peat around, just below and just above the track yielded similar dates, at the other sites peat from just below the track had so much the greater age as to support the field evidence that some erosion or wastage of the surface of the old peat had taken place before the track was laid upon it. We were now assured of the synchroneity of the major flooding horizon, and the curator of the Taunton Museum provided a sample of wood from the Shapwick Station dug-out boat, whose radiocarbon age of 2305 ± B.P. places the vessel in the Early Iron Age. This of course is younger than the wooden trackways we have been considering and accords with the reference of the boat to the *Cladium–Hypnum* peat of the first flooding episode. We were also able to date a small birch tree that had rooted in the *Sphagnum* peat just over the peat of this flooding episode: it was sample Q-30 and gave a date of 2240 ± 110 B.P. This is younger than the first flooding horizon but older than the second one, whose age is guaranteed by the typology of the Iron Age fibula found in it as around 1900 B.P.

By 1959 we had every reason to be satisfied with the concordance between the results yielded by the purely physical objective age

Fig. 47. Diagram shewing how the application of radiocarbon dating by 1960 had effectively confirmed earlier conclusions, based on pollen analysis, stratigraphy and archaeological evidence, as to the age of the prehistoric trackways then known. The age determinations of peat above and below the track support that of the wood from the track itself. The trackways fall into two age groups: the Late Bronze Age tracks are between about 900 and 500 b.c. in age; the Neolithic tracks are at least 2000 b.c., and are less strongly age-grouped than the younger series.

measurement and the conclusions derived by the coincidence of evidence from our earlier field and laboratory studies. This gave well-warranted hope of applying radiocarbon dating much more widely both within the Levels and further afield.

10

Geology of Levels and lakes: marine transgressions and lake settlements

Looking westwards to the Bristol Channel from Glastonbury Tor, along the Levels that lie between the Wedmore Ridge and the Polden Hills, it is quite evident that this is a submerged landscape in which the hill-slopes dip steeply down beneath the flat infilling of ancient peat cuttings and green pastures. Nor should this conclusion seem in the least surprising since at various great inlets on our western shores, such as Swansea Bay and Borth Bog, we have the clearest evidence of a great marine transgression produced by the rising ocean-levels during melting of the world's immense ice-sheets at the end of the latest Ice Age. In Swansea Bay our pollen analyses had made it clear by 1940 that the last 50 ft (15 m) or so of this great eustatic rise of ocean-level had taken place in the latter part of pollen zone VI, the termination of the Boreal period, thought possibly to date from about 5000 B.C. Here in the Somerset Levels deep borings had proved the bottoms of the drowned valleys to extend at least to 90 ft (27 m) below sea-level, and the type of stratigraphy recorded for our first section on Shapwick Heath (Fig. 15) proved to be repeated consistently everywhere in the peat lands of the Levels. Throughout the area the peat deposits were demonstrated to rest upon the surface of a soft blue-grey clay whose content of Foraminifera and of diatom shells shows that it formed in brackish water. The valleys were filled with this estuarine deposit right up to present-day sea-level and upon its extensive, flat, water-logged surface brackish *Phragmites* swamp developed, to progress through the normal sequence of automatic vegetational change to fresh-water fen, fen-carr and finally acidic ombrogenous raised bog. This sequence, repeated time and again throughout the Levels, was of course no more than we had already encountered at Borth Bog and just across the Bristol Channel in the great raised bog behind the coastal dunes of Swansea Bay. As at those sites so here, pollen analyses referred the base of the peat mires to an early stage of zone VIIa, the early Atlantic period.

In the East Anglian Fenland the black peat areas are separated from the Wash by a belt of silts and fine sands deposited under marine

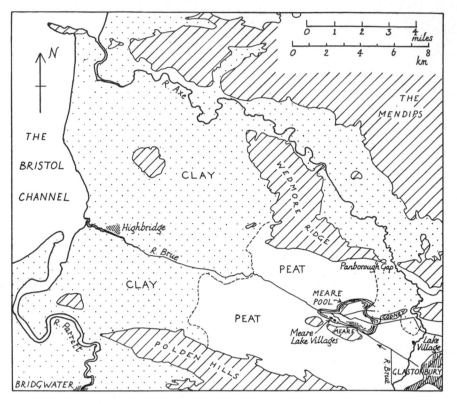

Fig. 48. Sketch-map of part of the Somerset Levels to shew the contrast between the clay-filled valleys traversed by the rivers Axe and Parrett and the valley between the Wedmore Ridge and the Polden Hills. This lacks any sign of a natural estuary and a coastal belt of clay has a hinterland covered with relict peat bogs, among which there persisted until historic time the large fresh-water lake, Meare Pool, that formerly extended to the Iron Age lake village site at Glastonbury.

conditions and extending to about present high-tide levels, so that we were very stimulated to discover that in Somerset also the seaward part of the Wedmore–Polden Levels consists of a belt of clay, 5 miles (8 km) or so in width and built up to about the level of 18 or 20 ft (5.5–6.0 m) OD. Faunal analysis again proved this very substantial deposit to have been marine in origin.

It was very fortunate for our investigation that in the early 1940s the Somerset Rivers Catchment Board began to cut a deep drainage channel straight through the clay belt from the peat land at Witchey Bridge, directly to the sea. We were given every facility to examine the many fine sections available (for a while at least) along its banks. Miss W. Abery made many measured sections, and here and there we were able to take sample series for pollen analysis and identification of macroscopic remains. By a laborious programme of levelled borings Roy Clapham and I compiled a section from Witchey Bridge to the Shapwick area, proving that the lower marine clay and the overlying raised bog peats persisted continuously in between: the Huntspill Cut sections extended the profile thus obtained, right to the coast, some 10 miles (16 km) in all. As may be seen from Fig. 49, the raised bog peats extended continuously seaward beneath the upper clay for about 2 miles (3 km), where they end in what seemed to us to be erosion channels where the tidal seas of the later transgression had met the edge of the bogs. The raised bog peats, now compressed by weight of the clay above them, were nevertheless still about 3 ft (1 m) in thickness and must have represented a long period of accumulation entirely free from marine influence. The top of the lower clay and the base of the old ombrogenous peat fell in pollen zone VIc close to the zone VIIa/VIIb boundary, but it was impossible by pollen analysis alone to date its upper surface. Fortunately here archaeological discoveries came to our aid, for along the Huntspill Cut at Newlands Rhyne, below the top clay and resting on the old peat surface, there were found traces of a Romano-British occupation that Bulleid had already reported from the adjacent peat surface. Similar remains occurred at this level elsewhere along the Cut and conformed to the discovery at Highbridge and elsewhere of datable Roman objects beneath or within the top clay. Since at other sites on the clay belt Romano-British occupation had been proved at the present surface, we were sure that, in a general sense at least, the second transgression began *and* ended within the restricted period of the Roman occupation.

What appears to follow from this is the important fact that right through the long period between the two marine transgressions the flat hinterland of the Levels was unaffected by any rise of sea-level. It was a long period covering the whole of pollen zones VIIa, VIIb and the opening of VIII, embracing the whole of the Neolithic and Bronze Ages and the pre-Roman Iron age, meanwhile presumably extending from about 5500 B.C. into the early part of the centuries A.D. Why this conclusion seemed of such interest, as I confess it still does, is the fact that in the East Anglian Fenland we had also been able to record the terminal stages of the great eustatic ocean rise, and that the last stages of

Fig. 49. Schematic profile from the Somerset coast, inland to the region of residual raised bogs. Thick ombrogenous peat extends seaward underneath the coastal clay belt for about 2 miles (3.2 km) and the Romano-British remains were found on its surface. The raised bogs rest on an estuarine clay that filled all the valleys to a little above present Ordnance Datum. Based on levelled borings and exposures of the deep Huntspill Cut, 1943.

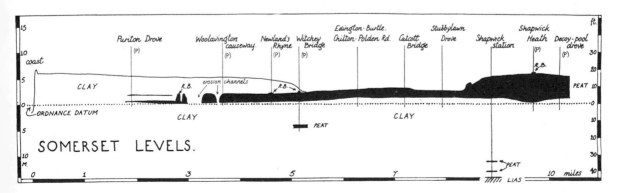

Fig. 50. Pollen diagram taken from samples accessible by cutting and deep boring at a site on the newly excavated Huntspill Cut, Somerset. The peat symbols employed are those of Fig. 38. The lower estuarine clay is succeeded by aquatic detritus peat, *Phragmites* reed peat (increasingly woody towards the top) and by a thick layer of *Sphagnum–Calluna* peat. On the surface of this were traces of a Romano-British salt-making industry, which was sealed in by the upper estuarine clay. Pollen zonation as for Shapwick Heath (Fig. 43) etc., places the end of the first marine transgression in zone VIIa and the onset of ombrogenous peat formation early in VIIb.

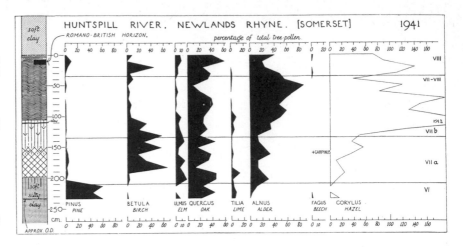

its geological history were a belt of marine deposits also of Romano-British age. However in the Fenland a major marine incursion had intervened between these two stages, that responsible for the Fen Clay, whose age was bracketed very clearly by a Neolithic horizon below and by an Early Bronze Age horizon above it. Accordingly there seemed no escape from the conclusion that whereas marine conditions had affected the Fenland drastically in this Bronze Age/Neolithic period, in Somerset there had been total freedom from marine influence right through this time. Given the coastal situation of both areas it seems hard to avoid the conclusion that during the period in question, the middle Flandrian, there has been a relative tilting downwards of the coast in the East Anglian Fenland. Certainly no world-wide upwards (eustatic) shift in ocean-level can have been responsible, and there is also other evidence around the coasts of the southern North Sea that suggests a tendency to recurrent down-warping.

Seeking a possible further analogy with the structure of the East Anglian Fenland, it seemed sensible to look inland of the coastal clay belt to see whether, in Somerset also, there might be evidence of a large natural estuary with raised banks of tidal origin which drainage had now exposed as the raised banks or levées of tidal rivers, i.e., in Fenland terms 'roddons', winding through the drained peat. However, neither on the ground nor in aerial photographs was there the least trace of such structures, nor, in correspondence with this, was there anything to correspond with the Fenland meres, fresh-water lakes held up behind the levées of the tidal streams. In fact the landward edge of the clay belt across the Wedmore–Polden Level was surprisingly straight and steep. It seemed likely in fact that throughout the Roman marine incursion this stretch of coastline had been unbroken by a natural estuary and that the

Fig. 51. Schematic comparison between the East Anglian Fenland and the Somerset Levels. Marine transgressive stages (clays and silts) shewn hatched, retrogressive stages (peat formation) shewn white, together with proved archaeological horizons. Between the early Neolithic and Roman time the Somerset peat bogs grew uninterrupted by marine incursion, but during this interval in the Fenland, marine transgression is represented by the 'Fen Clay'. Radiocarbon dates, (now written 'b.c.'), accord. Apparently, therefore, at this time the Fenland tilted downwards relatively to the British Channel area.

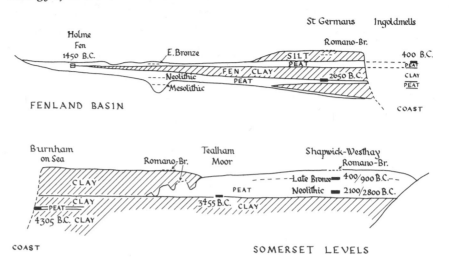

clay belt had been deposited up to and upon the margin of the pre-existing complex of raised bogs that occupied the flats. The present-day estuary of the River Brue, emerging through Highbridge, is a very minor waterway for which an artificial origin is likely.

It is entirely otherwise in those levels which lie just north of the Wedmore Ridge and in those to the south, leading into Sedgmoor behind the Polden Hills. Here respectively the rivers Axe and Parrett each has a large and active estuary where the stream displays its great natural meanders and, in the latter case, carries its burden of coastal shipping upstream to Bridgwater, just as in the Fenland the ports of Wisbech and Boston serve to link the agricultural hinterland with sea-borne transport. Borings and excavations alike disclose that the Roman marine transgression brought thick deposits of estuarine clay far up the Levels of both the Parrett and the Axe valleys. We resolved to trace the Roman clay as far up the Axe valley as possible in the hope eventually of making clear its relationships with the former large fresh-water lake of the Meare Pool, on whose shores were the Iron Age settlements that had first drawn us to the region, and with the Glastonbury Lake Village excavated with such outstanding success by Bulleid and St George Gray. We found that the Roman clay had deeply covered the pre-existing peat mires of the Axe Levels and then had squeezed through the narrow gap in the Wedmore Ridge between Panborough and Bleadney, south into the upper reaches of the Wedmore–Polden flats. Our attention had been first called to it here by a remarkable 'shark's skin' patterning in the aerial photographs south of the Panborough gap. Field sorties with the spade and borer proved this due to the presence of a clay layer just below the peaty surface, and it

Fig. 52. Sketch-map of the Meare Pool region shewing the approximate limits of the Romano-British estuarine clay intruding through the Panborough–Bleadney gap and south by Godney to Batch Farm. West of it is a landscape of exploited turbaries. Meare Pool is held up by a rock threshold at Westhay (WB 1,2); its historic extent is shewn here but earlier it extended eastward as far as the Glastonbury Lake Village. A line of borings links the raised bog margin in the west at PF 1–5 with the full extent of the lake (Fig. 53). The original River Brue flows north to join the River Axe; the existing River Brue is entirely artificial.

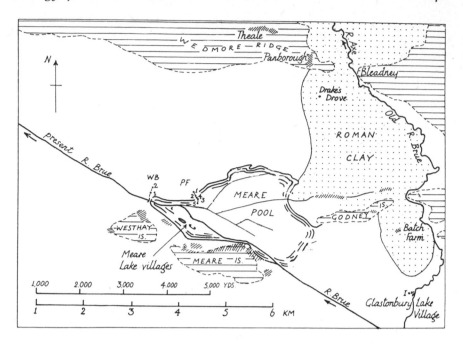

soon became apparent that the clay extended south to the small island of Godney, forming an area of fertile soil far more productive than the surrounding acidic peat, and fully exploited therefore for arable cultivation. The western edge of the estuarine clay tapered off alongside the Godney ridge, but extended far enough for us to shew that it entered the sequence of open-water deposits of the former Meare Pool, and furthermore, that at this point after the intervention of more organic lake muds, it was succeeded by the surface layers of fresh-water flood-clay that now indicate the site of the historic Meare Pool.

This was close to the limit of landward penetration of the Roman clay in the Wedmore–Polden flats, but we were able to shew that a thin spread continued eastward round the end of Godney Island, extending as what is now a slight rise or 'batch' to within about 200 m of the Glastonbury Lake Village site. Dr Macfadyen's analyses of Foraminifera from even this most inland extension of the Roman clay proved its estuarine origin quite conclusively. The evidence indeed supported Dr Bulleid's long-held view that in the whole river system from Glastonbury, beside the Glastonbury Lake Village and through the Godney and Panborough gaps, and along the Axe Levels we had the natural bed of the River Brue unable to find discharge directly westwards along the Wedmore–Polden flats. Whatever prevented this could well have been responsible for the creation of the Meare Pool and accordingly, through the period 1946–8, we spent a great deal of effort on field and laboratory investigations of the deposits of this historic lake and its close neighbourhood. The field work was often made exciting by the habit of keeping bulls along with the cows in the open pastures and one constantly had to estimate the defensive potential of a steel peat auger or one's ability (in gum-boots) to jump a 10-foot (3-m) rhyne. The detailed results of our field work were profuse and complex and here it will suffice just to outline the main conclusions we were able to reach.

(*a*) Right through the bed of the Pool, the soft blue clay of the great eustatic transgression was found close to present sea-level.

(*b*) Over a large part of the former lake, as at the Glastonbury Lake Village itself, lake muds had formed continuously from the end of the brackish-water stage early in pollen zone VIIa, through into the Iron Age of zone VIII: thus the lake had been in existence five or six thousand years before the Roman marine transgression.

(*c*) Rather more than a mile below the downstream exit of the lake a series of borings at right angles to the present artificial course of the Brue, and extending down to the lower estuarine clay, exhibited no trace whatever of any natural river channel that might have carried the outflow of the lake westwards to the sea: all that was encountered was the familiar peat sequence through from fresh-water fen and wood peat to deep deposits of raised bog overlaid shallowly on either side of the canalised river by its recent flood-clay.

(*d*) Where the Westhay Bridge carries the road across the present River Brue one may see Liassic limestone shallowly exposed in the river bed and a line of levelled bores carried across at this point to the Toll Gate House proved that indeed the lake outflow is constricted here: and no doubt in part this was what, along with the massive growth of raised bogs across the flats to seaward, caused the Meare Pool to form where it did.

(*e*) When we were carrying out our investigations, it has to be understood, there was no water in the great pool, but its historic outline was entirely apparent from the fact that it had been filled with a

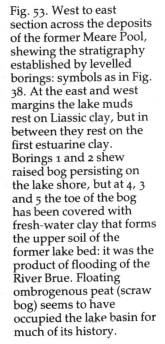

Fig. 53. West to east section across the deposits of the former Meare Pool, shewing the stratigraphy established by levelled borings: symbols as in Fig. 38. At the east and west margins the lake muds rest on Liassic clay, but in between they rest on the first estuarine clay. Borings 1 and 2 shew raised bog persisting on the lake shore, but at 4, 3 and 5 the toe of the bog has been covered with fresh-water clay that forms the upper soil of the former lake bed: it was the product of flooding of the River Brue. Floating ombrogenous peat (scraw bog) seems to have occupied the lake basin for much of its history.

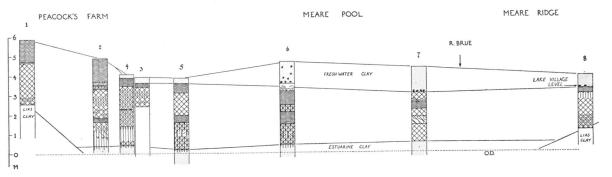

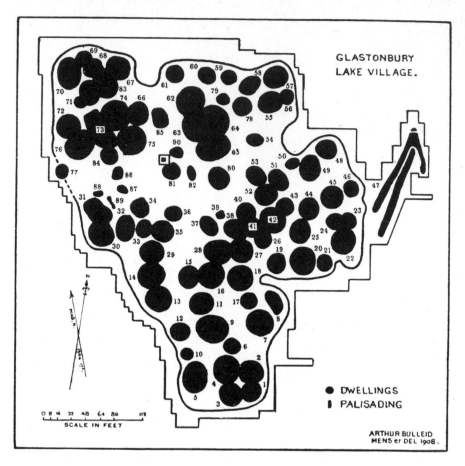

GLASTONBURY LAKE VILLAGE.

● DWELLINGS

▌ PALISADING

SCALE IN FEET

ARTHUR BULLEID
MENS et DEL 1908.

Fig. 54. Plan of the Glastonbury Lake Village drawn by Bulleid in 1908, shewing the separate dwellings, the surrounding palisade and, on the east, the causeway connecting the colony with the low upland (and river) to the east.

buff-coloured clay of about 1 m thickness, which was full of the shells of fresh-water molluscs. It is certainly a fresh-water flood-deposit associated with historic land clearance in the upper reaches of the river system and relates back to the time recalled by Dr Bulleid when the burden of flood-borne clay was so heavy and so predictable that it was regularly trapped in large settling tanks for subsequent spreading upon the mineral-deficient surfaces of the all-too-prevalent acidic bog land.

(*f*) We were still able to see, on the north-western shore of the former lake, how its deposits banked up against and partially overrode the untouched margins of old raised bogs: a levelled line of borings across the Pool from these margins to the Meare lake-village site opposite showed, beneath the continuous mantle of flood-clay, a deep and complex organic lake deposit.

(*g*) A great deal of the infilling of the Meare Pool beneath the blanket of clay was a remarkable and irregular admixture of two components, one of aquatic plants with sedges, bog-bean, pond-weeds and abundant algae, and the other of raised bog material, some fresh and undecayed but much of it decayed and blackened *Sphagnum–Calluna* peat such as could have been derived from erosion of adjacent older raised bog. It now seems likely that over much of its life-time the Meare Pool was occupied by what the Irish call 'scraw bog', a type of wet mire in which large floating mats of *Sphagnum–Calluna–Eriophorum* vegetation exist in a matrix of semi-solid submerged plant life. The whole is distinctly oligotrophic in character and the islands continually grow, sink and consolidate in the squelchy mass. Examples of such mires have been subsequently described both in the living state and from stratified deposits, but in the middle 1940s they had been scarcely recognised.

Thus, all in all, it seemed probable that the Meare Pool had originated early in pollen zone VIIa, upon the surface of the estuarine clay, and that the natural shape of the solid and drift geology of the valley had provided a barrier just seawards of Meare and Westhay which, aided by active growth of raised bogs, had closed off drainage of the flats to westward. The waters of the River Brue and associated streams had supplied the lake right through until historic time. Although overgrown by scraw bog and flanking raised bog, at the upstream end it remained an open lake and it was on the edge of this that the Glastonbury Lake Village was constructed and occupied between approximately 50 B.C. and 20 A.D.

When in 1946 and again in 1947 I went, fortified by instructions from Dr Bulleid, to the site of the lake village, I found it as he himself had done, upwards of sixty years earlier, recognisable only as a conglomeration of low circular mounds set in a pasture. It was whilst he had been a preclinical medical student at Bristol that he had found this site during vacation exploration, lake villages in mind, from his Glastonbury home. It was whilst still a medical student that he had made his first excavations and to see them on a Sunday stroll came a prosperous Australian sheep-rancher, formerly a native of Glastonbury, with his beautiful daughter. The attraction of the two young folk was instant and before the vacation ended their views were declared. Bulleid by now was determined to make his career in archaeology, then entirely without promise of employment, and the wise prospective father-in-law struck the bargain that if Arthur would return and qualify in medicine, he for his own part would undertake to bring back his daughter to England to marry her young man. The bargain was kept and Bulleid pursued his life as general practitioner in the Somersetshire coalfield, taking off what time could be spared for spells of archaeological research in the Levels. Doctor, archaeologist, geologist and artist he remained intensely cultivated, critical and human, utterly devoted to his wife but pretending

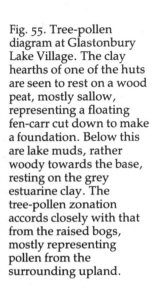

Fig. 55. Tree-pollen diagram at Glastonbury Lake Village. The clay hearths of one of the huts are seen to rest on a wood peat, mostly sallow, representing a floating fen-carr cut down to make a foundation. Below this are lake muds, rather woody towards the base, resting on the grey estuarine clay. The tree-pollen zonation accords closely with that from the raised bogs, mostly representing pollen from the surrounding upland.

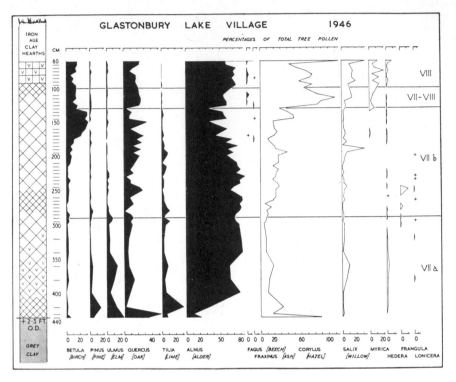

that it was she who had kept him from an ideal life of archaeological research. I have always counted it as one of the major benefits and pleasures of the Somerset investigations that they brought me into contact with so distinguished and lovable a man.

The borings made at the lake village site conformed to the account that Bulleid and Gray had published, establishing that it had been built entirely upon open-water detritus muds sitting above the surface of estuarine clay about 2.4 ft (0.75 m) OD. The transition from brackish to fresh-water conditions at that level was indicated by high values of the pollen of Chenopodiaceae, typical salt-marsh plants, and somewhat higher came maxima in the pollen of reeds, reed-mace and sedges that were the precursors of the open lake. Lake muds 10 ft (3 m) in thickness still remained, terminating above in a compressed wood peat representing the floating fen-wood of alder and sallow which Bulleid and Gray showed to have been felled *in situ* before deposition of the wooden framework of the hut-foundation and its successive floors of clay. The settlement consisted of between fifty and a hundred of such individual hut-mounds, all enclosed within a stout marginal palisading, and linked to the drier ground by a paved causeway. It was indeed a floating but anchored crannog settlement of the type several times excavated subsequently in Ireland, but the wealth of material retrieved from its pioneer excavation has been unmatched. Our borings had of course confirmed the difference in structural type between this offshore crannog and the later-excavated villages at Meare, where, on the margin of the same stretch of open water, the hut-mounds had been built upon the flank of the adjacent raised bog, again after introduction of a timber substructure. The lack of oxygen in all but the surface muds ensured the preservation of the vast amount of organic rejecta that found its way overboard into the waters of the lake or got trodden down into the soft earth. It is this that allowed Bulleid and Gray to make such dramatic and assured reconstruction of the life-style of the settlement and the quality of the local environment. It was a wealthy and versatile community farming the adjacent hill-slopes, herding domestic animals (especially sheep), growing cereals and the small celtic bean (*Vicia faba*), and upon the settlement site pursuing crafts such as weaving, potting, carpentry, basketry and, to judge from the abundance of sloe stones, also dyeing. To this variety they added the abundant natural resources of the region, not least those of the lake with its innumerable fish and wild-fowl, and creatures no longer wild in Britain such as the crane, the pelican and beaver. With such wealth of evidence emerging from his excavations, it is easy to sympathise with the desire of the young Arthur Bulleid to make his career in archaeology.

Fig. 56. Reconstruction suggested by E. K. Tratman of one of the square wooden houses of the Glastonbury Lake Village, using Bulleid's original records.

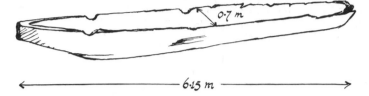

The outline Flandrian history of the Levels now presented has been based upon comparative studies of stratigraphy, archaeology and pollen analysis, together with specialist help of micropalaeontologists reporting upon fossil fauna and flora. As was the case with investigations of the upper peats and their contained trackways, this total information holds together in a self-consistent manner. None the less, although the recent end of the evolutionary story was adequately dated by Roman and Iron Age artefacts found numerously associated with the Roman clays and the lake-villages, the great marine inundation that marked the creation of the Levels by filling the valleys with estuarine clays, had been dated entirely by pollen analyses that indicated transition to fresh-water conditions early in pollen zone VIIa, the opening of the Atlantic climatic period, thought by comparison with the European mainland chronology to fall about 6000 B.C.

It will be imagined with what interest, therefore, as soon as the Cambridge radiocarbon dating laboratory had become effective, we sought to collect from the Levels samples for carbon dating that would represent beyond doubt the critical last stages of the marine transgression. These were found firstly on the foreshore at Burnham-on-Sea where at −15 ft (−4·5 m) OD a layer of *Phragmites* (reed) peat intervened in the uppermost layers of the blue estuarine clay. Collected in 1955 it belonged to pollen zone VIIa:

Q-134   Burnham-on-Sea   6262 ± 130 b.c.

In the same year at a deep drain cutting on Tealham Moor we were able to secure from a clean, freshly open section, samples representing the junction of the lower estuarine clay and the immediately overlying *Phragmites* peat close to present mean sea-level. Again the pollen zone proved to be VIIa and two continuous samples gave the following results:

Q-120   Tealham   5412 ± 130 b.c.
Q-126   Tealham (immediately above Q-120)   5620 ± 120 b.c.

It was satisfactory that these results should conform so well with one another (the Burnham sample was naturally older than those taken at the actual clay surface), and with the dating indicated by the tree-pollen zonation. The satisfaction was the greater since all the peat mires of the Levels were built upon the flat surface of the estuarine clay, so that in every sense we had an assured basis for all our future stratigraphic work.

11

Old peat and Neolithic culture

Whenever we probed into the ancient mires of the Levels or were able to examine exposures of its deposits we unfailingly found the same stratigraphic sequence. Everywhere the peat rested upon the flat estuarine clay whose upper surface was transformed some 7500 years ago, via a stage of salt-marsh vegetation, into brackish-water reed-beds. As the water freshened, the tolerant common reed, *Phragmites*, was joined or replaced by the giant sword sedge, *Cladium mariscus*, and in time, by the natural processes of vegetational succession, the thickening sedge-fields were invaded by bushes of sallow, alder and birch. Thus came into being a stage of extensive fen-woods easily recognisable by the abundant macroscopic remains of wood, bark, leaves and fruits, as well as by great frequencies of the appropriate tree-pollen types and spores of the marsh-fern that grows abundantly in such conditions. At this stage the ground-level emerged from the water-surface and its growing acidity began to be indicated by the invasion of *Sphagnum* moss, whose cushions continued the peat-building and acidifying processes so that before long the fen-wood floor was water-logged, its bushes and small trees suffocated for lack of oxygen and treeless raised bog took possession. Such bog was initially wet but came before long to be dominated by *Sphagnum, Calluna* and *Eriophorum*, with their characteristic calcifuge associates. This vegetational stage, when mature raised bogs occupied virtually the whole of the Levels, is clearly indicated stratigraphically by the old highly humified dark peat that has been so much valued throughout the centuries as fuel. This condition, broadly speaking, persisted until the first major flooding episode, which as we have seen, coincided with the ending of the Bronze Age and the building by its people of their numerous and elaborate wooden trackways.

Having, so to say, already dealt with the upper part of the stratigraphic sandwich, the younger peat formed during pollen zone VIII and in the archaeological periods from Late Bronze Age to Roman, and likewise with the lower part, the estuarine clays of late zone VIc and

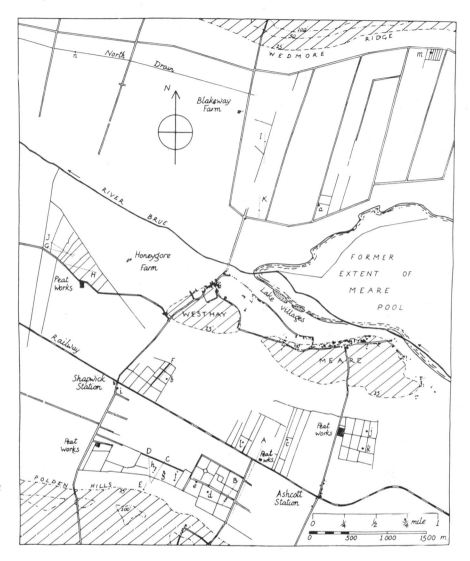

Fig. 58. Artefacts known in 1963. Neolithic trackways were G, Honeygore; H, Abbot's; I, Blakeway Farm; J, Honeycat. Other Neolithic artefacts were e, Graig Lwyd stone axe; j, Meare Health pottery; l, and k, yew wood bows.

early VIIa, we are left to consider the beds of dark humified acid peat consistently layered between them in the manner just described: the meat in the sandwich.

I had been introduced to the stratigraphic sequence as early as 1935 by our first boring on Shapwick Heath – and found it immediately repeated when, in the very next year I was taken to see the site on Meare Heath where broken sherds of a Neolithic pot of Peterborough ware had just been found resting close to the base of the old *Sphagnum–Calluna–Eriophorum* peat. At the top of the reed-swamp peat, only 20 cm lower, high values of birch pollen disclosed the local presence of the transi-

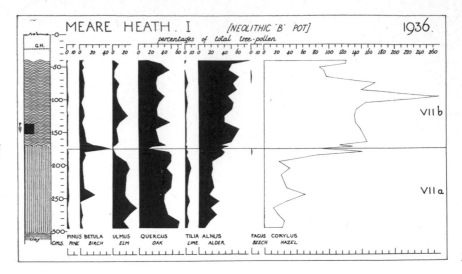

Fig. 59. Site of discovery of Neolithic 'B' pottery on Meare Heath in 1936. The pot came from just above the base of the old humified *Sphagnum–Calluna* peat and just above the pollen zone boundary VIIa to VIIb.

tional fen-woods, though actual wood peat was not observed.

In the years immediately following, because peat digging in the Shapwick–Meare region was extensively concerned with the upper 2 m of peat, we found ourselves, as has been recounted, very busily involved with all the phenomena revealed from the archives of the younger peat, and nothing older than Middle Bronze Age came to our notice. None the less the tree-pollen diagrams made at many sites extended down to the basal clay and were remarkably similar to one another. Each of them appeared to cover the whole of pollen zones VIIa and VIIb, with the contact very close to the base of the old *Sphagnum–Calluna–Eriophorum* peat. These zones had been defined in 1941 as part of the zonation system for the whole of England and Wales; the feature distinguishing them was the abundance of elm, lime and, to a lesser degree, pine pollen in the earlier zone, together with the abruptness of diminution of these values at the zone boundary. When the zoning was proposed it was based solely upon the consistency with which it recurred: it was only later that the European pollen analysts indepen-dently adopted the same zone criteria, and shortly afterwards suggested possible explanation of the phenomenon in vegetational terms. The Somerset pollen diagrams were all characterised by extremely high frequencies of elm pollen in zone VIIa, often persisting at more than 20 per cent of the total tree-pollen, and one is forced to accept a reconstruc-tion of the woodland cover of the surrounding hills, the Poldens, Wedmore Ridge and Mendip, as containing a very high proportion of this tree and of lime (which is much under-represented in its pollen yield) within the mixed oak woodlands. Suffice it for the moment that the VIIa/VIIb zone boundary apparently offered a consistent time-horizon within the old peat deposits.

We next encountered the Neolithic culture in 1944 in the shape of the highly distinctive wooden trackway at Blakeway Farm, between West-hay and Mudgley. We were told by the helpful farmer of a track running north–south 'straight as a die' and made of straight stems of

Plate 52. The Blakeway Farm Neolithic trackway as excavated in 1945, looking north. It shews the butt ends of one faggot of straight hazel rods resting upon the tapering ends of the next one causing a fracture zone.

'nut-wood'. Since no surface trace remained, and we had only a general idea of location, Roy Clapham and I, in determined mood, sank a row of thirty-six borings at 1-ft (0.3-m) intervals across the presumed line of the track and eventually we made six borings, only a few inches apart, each containing at the same depth the typical clear yellow uniform wood of hazel lying horizontally. We were certain this was indeed the track and later in the year were able to excavate a substantial length of it. It proved to lie about 3 ft (1 m) below the peat surface and to consist of a single row of parallel and very straight rods of hazel laid without cross-ties, and with only a few small oblique stakes at the margins, to make a track only about 2 ft (61 cm) wide. It had been built upon a thin layer of transverse *Calluna* brush, and consisted of a series of successive faggots of hazel, each of about twenty rods placed closely side by side and arranged so that the butts of each faggot somewhat overlapped the thinner apices of the next. The longest hazel pole was 13 ft (*c.* 4.0 m) long but many more approached these dimensions and all were astonishingly straight and free from branching (see Fig. 60). It was entirely clear that natural woodland never grew hazel of this kind, and when we cut and compared the ring-structure of the hazel poles (between 0.9 and 1.5 in (23 and 38 mm) in diameter at the base) it seemed certain that we were dealing with the product of a coppiced woodland. So much we concluded and published at once: we remained uncertain of the age of this product of early sylviculture, although we pointed out that the axe-cuts on the rods had left a concave surface and a curved blade-mark suggestive of the use of a small axe of considerable thickness.

The stratigraphy at the trackway site conformed to the pattern already described, with the trackway about 1 ft (30 cm) above the well-developed layer of fen-wood peat, a layer that seemed to have formed over the time of the pollen zone VIIa/VIIb boundary.

We were unwilling, I suppose, to conjecture that in this unexpected way we had fallen upon (almost tripped over) a Neolithic trackway, and our uncertainty was heightened by the fact that the raised bog peat in this part of the moors did not show the usual decisive twofold division

Fig. 60. Scale plan of the excavated Blakeway Farm track, shewing spread faggots of hazel poles laid consecutively to make a trackway. Occasional slender cross-bearers and oblique stakes were also used to improve stability and prevent lateral spreading. The straightness and the ring-structure of the hazel poles indicate clearly that they were coppice-grown, even though the age is Neolithic, with a radiocarbon age subsequently determined as 2510 ± 130 b.c.

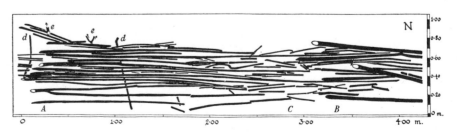

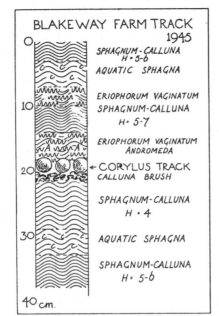

Fig. 61. Blakeway Farm track: stratigraphy revealed by dissection of a peat monolith crossing the track. Layers of aquatic *Sphagnum* peat occur above and below the track and seem to suggest that locally at least it was laid on rather wet bog. Symbols as in Fig. 38.

into upper and lower peat. What was worse, we encountered to the south, at Toll Gate House, a trackway precisely aligned with the Blakeway Farm track and of similar construction. This was accessible in an open peat face. As we have seen, however, the pollen analyses, peat stratigraphy and ultimately the radiocarbon date confirmed that the Toll Gate House track could not be other than of the Late Bronze Age.

To the dilemma of dating the Blakeway Farm track and its dependent sylvan culture, radiocarbon dating eventually in 1960 made its decisive resolution. The axe-sharpened ends of hazel stems that had been displayed in a museum jar since 1944, were baked free from the preservative alcohol, combusted and assayed, to yield the date:

Q-460   Blakeway Farm track   2510 ± 130 b.c.

This is Neolithic by any standard, explaining the absence of lateral cusp-marks on the axe cuts and amply confirming the astonishingly early existence of hazel coppice or closely similar anthropogenous vegetation on the hill-slopes of the region.

In 1947 we were notified of a trackway apparently running east–west between Westhay and Catcott Burtle: we named it 'Honeygore track' from the adjacent farm with that distinctive title, and having verified it at once, my research student, Joe Jennings, and I returned in the following year to excavate, record and sample the site. It proved to be a trackway of substantial birch timbers, up to 3.5 or 4 in (9–10 cm) in

Plate 53. Honeygore Neolithic track, Westhay, at an early stage of excavation, 1948. All the main timbers are of birch laid longitudinally. The radiocarbon age was later determined as 2810 ± 130 b.c.

diameter, laid longitudinally and closely beside one another, covered with a considerable thickness of brushwood (now reduced to trash), with frequent transverse sticks below the main timbers and many stakes pinning the track down, especially at the margins. The whole structure was about 3.5 ft (1.2 m) in width, and was specially notable in that an overturned alder tree had been incorporated, when it was growing, into the track structure. This was a feature we encountered again in 1959 when we relocated the track to secure samples for radiocarbon dating. In 1948 our opportunities for dating the trackway were less direct. Nevertheless we had three important indices to go upon. The peat cutting exposure was supplemented by boring and together the results confirmed the standard stratigraphic pattern, with the trackway sitting directly upon a coarse detritus wood peat, indicative of the fen-woods in which the incorporated alder was growing: below there was *Phragmites* reed peat with *Cladium* and bog-bean, giving place at the bottom to the usual soft blue clay grown through by reed. Pollen analyses from below the track made it probable that the reed-swamp and fen-wood peats had formed at the same time here as in the Shapwick area, i.e. through zone VIIa, and finally, the main birch timbers found both in 1948 and 1959 showed the marks of thick-bladed axes with a curved edge, but without lateral cusps. By hindsight it seems very clear that the Honeygore track was Neolithic, but we remained uncommitted to this view until we had the radiocarbon results of our 1959 samples, viz.

Q-431   Honeygore track   2810 ± 130 b.c.

It could not have been more conclusive.

Despite recovery of these trackways we had to wait until 1957 for our most direct and decisive evidence linking the Neolithic culture with the other indices of evolution in the Levels. It came with the discovery by peat cutters on Shapwick Heath of a polished Neolithic stone axe, which, when sliced and examined microscopically proved to be one of a number of such axes coming from the well-known group of axe-factories at Graig Llwyd in Caernarvonshire (Gwynedd). We were able to determine the familiar peat-stratigraphic sequence as given here:

0–48   Reworked peat of former digging
48–130   Highly humified dark-brown *Sphagnum–Calluna–Eriophorum* peat, containing the axe at about 120 cm

Plate 54. Polished Neolithic stone axe found on Shapwick Heath in 1957, here seen in two views. Petrographic analysis proved it to have come from the well-known Neolithic axe-factory site of Graig Llwyd mountain in North Wales. The tool was found near the base of the old highly humified *Sphagnum–Calluna* peat slightly above a birch wood-layer. It was from this transitional horizon that a sample was taken for radiocarbon dating.

130–154   Laminated *Sphagnum* peat with abundant birch wood
154–264   Coarse detritus reed-swamp peat with *Cladium* in the lower part
268–287   *Phragmites* peat with large wood *in situ*
287–   Soft blue clay with *Phragmites*

Pollen analyses already made from three sites nearby all confirmed that the base of the old humified ombrogenous peat must lie at the beginning of zone VIIb, and no fresh analyses were undertaken. Stratigraphic level and pollen zonation were thus very similar to those of the 1936 discovery of the sherds of a Peterborough ware pot on Meare Heath. Now, however, we were able to draw upon the resources of radiocarbon dating in substantiation. The dated samples, collected two years after the discovery, were highly trustworthy as to provenance and by 1960 gave these results:

> Q-430   Shapwick Heath Neolithic axe site: wood from base of old *Sphagnum–Calluna* peat, at level of Graig Llwyd axe.   2580 ± 130 b.c.
> Q-423   Shapwick Heath Neolithic axe site: upper surface of the estuarine clay beneath the raised bog.   3560 ± 120 b.c.

The correspondence of the one date with that of the Blakeway Farm track, and the substantially greater age of the other is very satisfactory and strongly confirmatory of the general correlation scheme. By 1960 we had also added two further radiocarbon dates for the base of the old ombrogenous peat on Shapwick Heath. Samples Q-115 and 116, they yielded respectively ages of 2060 ± 120 and 1630 ± 120 b.c., both so much younger than Q-430 as to suggest, as indeed was likely, some local variation in the time of transformation from fen-woods to acid bog vegetation.

The pottery and the polished axe were themselves evidently artefacts of Neolithic age, but now there came to our notice two important objects whose cultural provenance it was possible to deduce only through reference to the conditions of discovery and radiocarbon dating of the associated peat. They were found, the one in February 1961 and the other in June of the same year, respectively in peat diggings on Ashcott Heath and about 1 mile (1.6 km) away on Meare Heath, and oddly enough each was half of a long-bow, made of yew wood (*Taxus baccata*) just as were the bows used at Crécy and Agincourt so much later. Thanks to the strong interest our researches had aroused in the area and a clear appreciation of what was needed, in each instance the half-bow was removed and conveyed to us along with a sufficient block of the peat on which it rested. From this we were able to take unexceptionable samples both for pollen analysis and for radiocarbon dating, whilst the objects, still water-saturated, were available at once for preservation by the *carbowax* process in the Cambridge Museum of Archaeology and Anthropology. An early visit to the two sites confirmed the stratigraphic provenance of the finds, i.e. within the lowest part of the old highly humified *Sphagnum–Calluna–Eriophorum* peat. The tree-pollen analyses

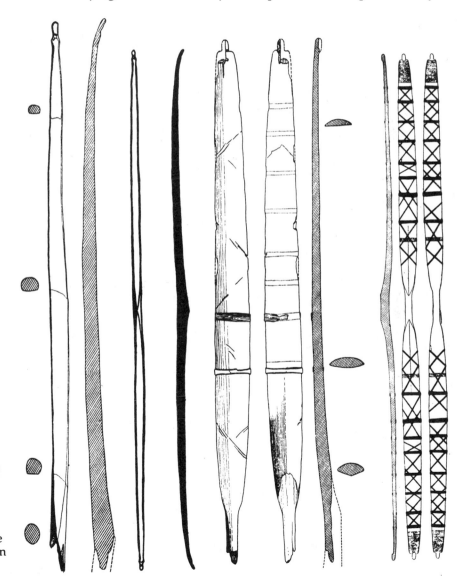

Fig. 62. Drawings of the two Neolithic yew bows found in the Somerset peat bogs: on the left the half-stave from Ashcott, on the right that from Meare Heath. Professor Clark's drawings shew face- and side-views for each half-stave together with his reconstruction of the appearance of each whole weapon. There are also measured cross-sections of each bow. Note the remarkable cross-binding of thongs on the Meare bow.

placed the finds within the lower part of zone VIIb, somewhat above the highly characteristic 'elm decline', at a time when the non-tree pollen showed indications of forest clearance in the presence of such 'weed' species as ribwort plantain (*Plantago lanceolata*), docks (*Rumex* spp.), goosefoot (Chenopodiaceae) and grasses (Gramineae). These strong indications of probable Neolithic age were vividly confirmed by the radiocarbon dates:

Q-598   Ashcott Heath long-bow   2665 ± 120 b.c.
Q-646   Meare Heath flat-staved bow   2690 ± 120 b.c.

These results were surprisingly congruent with that for the Neolithic axe.

Our two bows, generously and wisely presented to the Cambridge Museum of Archaeology and Anthropology, proved of great archaeological interest and indeed were the stimulus that provoked the writing by Professor Grahame Clark of a classic paper taking in the prehistory of archery in the whole of north-western Europe.

The Ashcott bow was closely similar in length and style to the mediaeval English long-bow: though broken at the hand-grip it could be estimated as having been just below 2.0 m long; it had a well-defined shoulder-notch for the bow-string, was half-round in section of the stave and oval in the hand-grip. In comparison with other known prehistoric bows the half-round limb was particularly thick, almost as thick as wide.

The Meare bow was of an outstanding type. Of overall length about 190 cm, the stave was remarkably wide, flat and thin, and of lanceolate shape, tapering gently to the terminal notch and more abruptly to the nearly cylindrical hand-grip. A most unusual feature was that there remained on the surface of the stave, portions of transverse binding thongs of hide or leather about 1 cm wide. These bands, originally some eight in number, were supplemented by diagonal criss-cross banding of narrower strips, and were themselves decorated by parallel cuts. One can only surmise that this webbing was intended to stiffen and safeguard the thin blade of the limb from damage, especially from marginal splitting. It must have been an exceptionally prestigious weapon, and the yew-wood facsimile made by the chief assistant of the Museum proved also to be a fully effective one, as judged by its performance when introduced to a national toxophilite meeting in 1963.

It will be apparent that by 1960 or thereabouts radiocarbon dating had become a final Court of Appeal on all issues concerning the ages of the ancient phenomena of the Levels and that it had marvellously confirmed the validity of the general correlation scheme independently built up beforehand. From then on there has been a tendency simply to rely upon the radiocarbon assay alone, a practice that is attended with some hazard since a number of potential errors lie in wait for the careless user, such as the contamination of the organic sample by younger material grown down from above as rootlets or passed down as urea or humic material in solution or, for that matter, by older material such as coal or

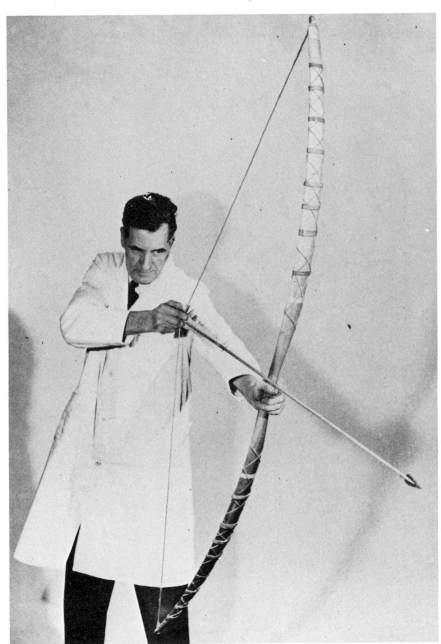

Plate 55. The chief assistant of the Cambridge Museum of Archaeology and Ethnology proceeds to draw a reconstructed copy he has made of the Meare Heath bow. It proved very serviceable in practice.

charcoal washed into the contemporary sample. Above all it is essential that the sample relates unequivocally to the episode or event it is desired to date, that it represents fixation of atmospheric carbon dioxide precisely during the event in the way that outer rings of tree stems do and that the inner rings do not. A multiplicity of circumstances must be taken into account, and it is highly to be desired that sampling in the field should be done under personal direction of the laboratory scientist. There remain many issues where he can take avoiding or corrective action, especially in choice of standards that are unaffected by atomic bomb effects in the atmosphere, or contamination of the air and biosphere by the products of burning fossil fuel. I say nothing of the problems of reducing radiocarbon years to the secular calendar, since for our present purposes it suffices to establish a chronology in 'radiocarbon years', a scale that only diverges largely from the solar one as the dates become older than about 2000 years before the present.

With our own radiocarbon dating laboratory in Cambridge and its staff closely involved in all our field investigations we were indeed fortunate and it was easy to interject the dating of new discoveries into our programme as they occurred. One such was the unearthing of yet another wooden trackway, apparently running between the two low islands of Burtle Beds, at Catcott Burtle and Honeygore Farm, the basis of our fanciful name, the 'Honeycat track'. This, excavated in 1959, proved to be a feature of slighter structure than the nearby Honeygore track, with longitudinal main timbers of birch laid, as in the Blakeway Farm track, in the form of successive single faggots. About 4 to 5 ft (1.2 to 1.5 m) in width the track incorporated a good deal of brushwood. The longitudinal timbers rested upon transverse bearers and again we had the phenomenon of a large stool of alder, growing on the spot, having been built into the fabric of the trackway. The stratigraphic sequence was shewn to be broadly similar to that already recorded for the Honeygore track, with the trackway again immediately above reed-swamp peat containing local wood. This indication of Neolithic age was confirmed by three radiocarbon assays:

Q-320   Honeycat track: wood of track, coll. Jan. 1959
       2115 ± 130 b.c.
Q-427   Honeycat track: wood of track, coll. Sept. 1959
       2376 ± 130 b.c.
Q-429   Honeycat track: wood of alder incorporated in track
       2265 ± 130 b.c.

It was by now entirely apparent that we had to do with two distinct groups of wooden trackways in the Levels. Whereas it seemed that the

Late Bronze Age tracks had been constructed in response to a climatic worsening that had flooded the surface of the raised bogs, submerging them largely in wet sedge-fen or similar communities, the Neolithic trackways had been built so much earlier that ombrogenous bog formation was only just beginning with the establishment of wet cushions of *Sphagnum* moss on the quaking surface of the prevalent fen-woods of birch and alder. How might such an environmental change be envisaged as having stimulated Neolithic man to link the low islands by lengthy and sometimes robust trackways? It can hardly be advanced that here was the onset of flooding of surfaces hitherto easily passable to him, and one is tempted to conjecture rather that, with the consolidation of a floating carpet of fen-woods above the reed and sedge-fens, for the first time it had now become possible to construct linking, if fragile, wooden pathways from one upland to another, particularly, if the field evidence is to be believed, from one sand batch to another. We trust that the future may yield us more evidence: what however seems already apparent from radiocarbon dates already available, is that the Neolithic tracks so far assayed cover a good time-span, as if the development of the fen-woods through the area was only very broadly synchronous, not at all an unlikely thing.

Had we been more aware of this kind of origin of the early trackways we could have been spared the failure of our attempt to find the earliest-recorded of all of them, the so-called 'Abbot's track'. This was the trackway discovered in 1834, demonstrated by C. W. Dymond in 1864 and excavated and recorded by him in 1873. It was seen later by Bulleid and had been set down on the Ordnance Survey maps, so that we supposed the line of it to be well known. However, several attempts having failed to find sections of it in recently opened peat faces, Roy Clapham and I began in 1942 (as later at Blakeway Farm) to put down

Fig. 63. The Abbot's track, known since 1834, was described by C. W. Dymond in 1873 and 1880, and from his drawings and text the tentative reconstruction shewn here was made (1960). This conception of it as made of transversely set, split timbers, held down by two lines of stringers and vertical lateral pegs, has not been greatly altered by more recent exposures of the track. Samples taken in 1961 were given a radiocarbon age of 2660 ± 120 b.c.

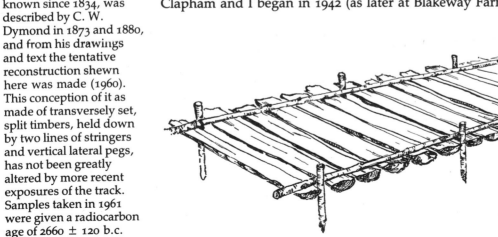

borings in fairly close order. These established the expected peat stratigraphy, but no sign of the track and our search was abandoned. I am sure that we concentrated on the false expectation that we were about to find a Late Bronze Age track at a flooding horizon fairly evident in the upper peat, and we failed to make the imaginative jump that would have told us that not *all* the frequent timber encountered at the junction of the swamp peat and the acid bog peat was attributable to the local fen-woods as we then supposed. Such are the dangers of preconceived ideas.

It was not until 1961 that the opening of pasture towards Westhay for further peat cutting was in fact to re-expose the Abbot's track close to the line already mapped for it. It was then found to lie in the stratigraphic situation of the other tracks of the region already proved to be Neolithic. This was now confirmed by a radiocarbon age determination:

Q-647   Abbot's track   2660 ± 120 b.c.

This result fits harmoniously into the general correlation scheme covering the older peat and the contained Neolithic culture, and it closes effectively the account of my own active involvement with field studies in the Somerset Levels, a consequence of heavy duties with such chair-borne administration as that amply provided by the Cambridge Professorship of Botany and Presidency of the impending International Botanical Congress in Edinburgh. Not that field studies in the derelict bogs ceased: far from it. From 1964 these have been pursued with enormous vigour and effectiveness by teams of students and research workers directed primarily from the Cambridge University Department of Archaeology: their results since 1975 have appeared in a series of fascinating issues of 'Somerset Levels Papers' edited by Dr John M. Coles, wonderfully expanding our knowledge, particularly of the detailed construction and building processes of the Neolithic trackways and of the culture responsible for them. Happily these results are nowhere substantially in conflict with the earlier results which I have been concerned to summarise here.

It has suited our purpose to have recounted the story of our own discoveries in the Levels broadly in the time-sequence in which they were made. Not only has this made for greater ease of telling but it suited the progress of technical means of research, and the alteration in the manner and extent of peat extraction. Thus not until 1952 did we begin to build a radiocarbon apparatus, nor was it yielding many results until 1959: prior to this time we relied entirely upon correlation of pollen analysis, peat stratigraphy and archaeological typology to establish the relative ages of events recorded in the deposits of the Levels. Then again, when we began work in the Shapwick area and for the next thirty years or so, peat cutting was largely being done by hand and was restricted to the upper 6 ft (2 m) of the bogs, so that inevitably it was the younger peat with its content of Bronze Age, Iron Age and Roman artefacts that was mostly displayed to us. Only later, as the younger

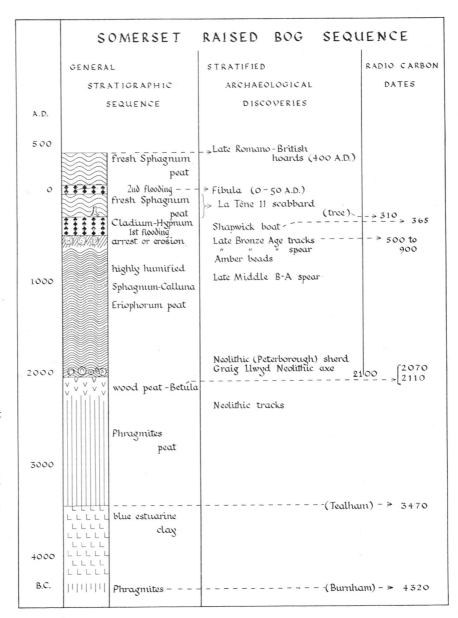

Fig. 64. Correlation table made in 1960 for the Shapwick–Meare–West-hay region of the Somerset Levels, shewing the relationship between bog stratigraphy, flooding horizons, pollen zonation, trackway construction and archaeological cultures. In this diagram it was not yet possible to include radiocarbon dating of the Neolithic trackways although their level is clear. All radiocarbon dates given here now are properly written 'b.c.'.

peat gave out and cutting returned to areas already exploited in the past and left aside, were the deeper peats subjected to renewed extraction. Thence it followed inevitably that we were concerned with the older acid peat with its contained Neolithic artefacts, and especially its fine wooden trackways.

Because of this progression in the field studies we were able to sketch our findings in three chapters each summarising our conclusion to a given date. In Chapter 9 we were able to describe the situation as it had evolved by 1951, when, with pollen analysis, peat stratigraphy and archaeological discovery we had reached a consistent explanation covering events from the Late Bronze Age. Chapter 10 represents the analytic stage of about 1955 when the earliest radiocarbon dating results could be attached to those of earlier techniques to interpret the geological evolution of the Levels. Thirdly, we have the situation as seen in 1961 when a series of important Neolithic discoveries had been made, the stratigraphy of the middle peats analysed and the whole checked into its chronological framework by a sufficient number of critically chosen radiocarbon dates, i.e. Chapter 11.

12

Disforestation and agriculture

It often appears that when some significant scientific advance is made one looks in vain for any particular individual or school as being the exclusive agent responsible for it. This was certainly the case in the origin of ecology as a scientific discipline: it arose independently about the beginning of this century, in widely separated parts of the world. It was as if the concept had been maturing with the development and interlocking of physical and biological sciences, so that it was spontaneously recognised in many quarters.

A similar instance arose in the development of pollen analysis during the early 1940s. In its early stages this method had been limited to the study of the pollen of forest trees and shrubs, and it had allowed us to build up a picture of the natural forests of central and northern Europe as they migrated across the continent under the influence of the changing climate that replaced the rigours of the last Ice Age. But with improving microscopy and determination to fill in the detail of past ecological conditions, attention moved to the recognition of more and more of the sub-fossil types of spores and pollen grains produced by the herbaceous plants and low-growing shrubs, so that they appeared increasingly in the pollen counts and in the pollen diagrams. This was the situation in 1940 when we took for analysis a series of samples through the deep open-water muds of a former lake, Hockham Mere, that lies at the northern edge of the big stretch of East Anglian heath known as the Breckland. Ecologists were uncertain why, instead of deciduous woodland, so much of the landscape was covered with *Calluna* heath, bracken, sand-sedge or open, heavily grazed grassland. It was attributed sometimes to the low regional rainfall, sometimes to the porous infertile soils and sometimes to the intensity of rabbit-grazing, but this was all quite conjectural. It soon appeared that the Hockham Mere deposits provided a continuous history of the regional vegetation from Late-glacial time onwards and that during the climatic amelioration of the Boreal and Atlantic periods, mixed oak forest was dominant in the region (Fig. 4): correspondingly, pollen of herbaceous or heath com-

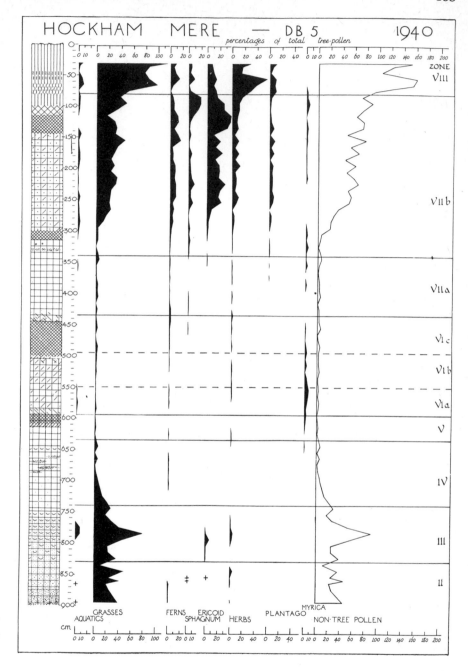

Fig. 65. Non-tree pollen diagram through the sediments of an East Anglian lake, Hockham Mere, on the edges of the heath-covered Breckland. The extremely low frequency of non-tree pollen in the middle Flandrian (zone VIIa) contrasts sharply with the high values thereafter. This was the earliest definitive evidence of the origins of Breckland in prehistoric woodland clearance. (Tree-pollen for this site is shewn in Fig. 4.)

munities was extremely infrequent. All this altered in the uppermost 3 m or so of the lake deposits: above this level there were maintained high values for grass pollen, heather pollen, fern spores (including bracken) and miscellaneous herbs, one of which, though highly typical and frequent, we had to be content to label 'X'. It seemed hard to resist the conclusion that these sustained high values of non-tree pollen represented the past dominants of the characteristic Breckland heaths. They had originated at a very recognisable horizon in the tree-pollen diagram, when a sharp fall in the percentage of elm pollen was the prime indicator of the Atlantic/Sub-boreal climatic boundary. Fuller comprehension waited upon events in Denmark, unknown to us until 1944 when a scientific offprint arrived via Sweden from Johannes Iversen at the Geological Survey in Copenhagen. This was a paper of the greatest ecological and archaeological interest entitled 'Land occupation in Denmark's Stone Age', and published in 1941. In a bog grown above a small lake, Iversen found evidence that there had been prehistoric settlement at a time well defined by the position of the Baltic shore-line and by the tree-pollen, as at the Atlantic/Sub-boreal contact: not only was there abundant charcoal in samples from this level but also a bone of a cow was found. At this horizon the total tree-pollen sharply decreased in its absolute frequency, the pollen of the oak-forest domi-nants also diminished, whilst that of birch, alder and hazel correspon-dingly increased. Together with these light-seeded and easily estab-lished trees, considerable amounts of the pollen of herbaceous plants were now encountered, very notably the large spherical grass pollen types attributable to cereals (barley and wheat), smaller non-cereal grass pollen, Chenopodiaceae (the family of goosefoot, fat-hen, etc.), *Artemisia* (wormwood) and finally *Plantago* of two species, *P. lanceolata* and *P. major*.

Comparable evidence came from other sites of a similar oscillation in vegetational history, one that it was impossible to interpret as climatical-ly caused. The picture of events plausibly reconstructed by Iversen was of the sudden arrival of Neolithic farmer folk who cut and burned the primeval forest, not in any great conflagration but in small clearances with controlled burning as still carried out in Finnish Carelia. Crops were taken of early races of cultivated wheat and barley sown in the ashes, and when the repetition of cropping had soon exhausted the soil, the clearance would have been abandoned to recolonisation by the more mobile woody plants, the early growth probably subject to browsing of its young leafage by cattle. In due course, as the pollen analyses shewed, the tall forest re-established itself, although the elm failed to reach its former frequency.

The word employed by Iversen for this temporary forest clearance by Neolithic man was *Landnam*, or 'land-take', perhaps better represented in English by the place name I used as a boy to see on the Sheffield trams

Plate 56. Dr J. Troels-Smith demonstrating in the woodland at Draved, how an actual Neolithic axe can be used to fell standing trees. This alder, 15 in (38 cm) across, was cut through in 15 minutes.

outward bound from the city centre, namely 'Intake'. No doubt the East Anglian term 'Breck' had similar significance, a pleasing thought since it was easy to see in my Hockham Mere analyses evidence of a remarkably similar event. What immediately struck me were the illustrations of pollen grains of the plantains, one of which, the ribwort plantain (*Plantago lanceolata*), I immediately confirmed as the type 'X' grain so consistently present in the Hockham diagram. This plant, in Danish *Vejbred*, is entirely restricted to low open vegetation, especially waysides, and must certainly indicate interference with the forest dominance. Perhaps it will be recalled that according to Fennimore Cooper, the boy's story-writer of my youth, the plantain (presumably the broad-leaved *P. major*) was called by the North American Indians 'the white man's footprint'.

There seemed, therefore, little doubt that at our site in the Breckland, that part of Britain most heavily occupied by Neolithic man, we had the evidence that sustained overlapping clearances of the original closed woodland had created the diversity of the present heath landscape. Greater probability still, followed from very similar pollen sequences taken in the extensive sandy heath region of Jutland, and it is a conclusion not seriously questioned since its publication in 1944. It was not long after this that Iversen and Dr J. Troels-Smith, a colleague from the Danish National Museum, set up an experiment in the nearly natural forest of Draved in Jutland, to reproduce the process of woodland clearance and cultivation as they thought it had been conducted.

Fig. 66. Type of haft employed for polished Neolithic axes in tree-felling at Draved, copied from an original found in a Danish peat bog.

Using actual Neolithic polished flint axes, hafted in the manner indicated by bog-finds of the wooden helve, they felled the alder and killed the oaks by ring-barking. The cut wood was burned on the ground in the spring and crops of the Neolithic wheats *Triticum monococcum* and *T. dicoccum* were raised in the ash. It was surprising to see how a flint axe, properly wielded, could fell a 15 in (38 cm) diameter alder within 15 minutes, and it was astonishing to see how by some means the broad-leaved plantain at once appeared within the cereal crop.

Evident as it had now become that the clearances of agriculture had henceforward to be given great significance in vegetational history, the role of climate control had still to be reckoned with and Iversen pointed to the decrease in pollen of elm and linden, as well as the frost-sensitive ivy, at the Atlantic/Sub-boreal boundary as perhaps indicative of this. At this stage we were engaged on analysis of the close pollen series from the Somerset site of Decoy Pool Wood on Shapwick Heath, where the sequence extended from the lower highly humified dark peat through the upper peat with its two flooding horizons, the lower of which

Plate 57. Crop of Neolithic wheat grown in a clearing in the forest at Draved, Jutland, after felling with stone axes and burning the brush. Dr Iversen points to the plantain (*Plantago major*) that has entered fortuitously during their husbandry.

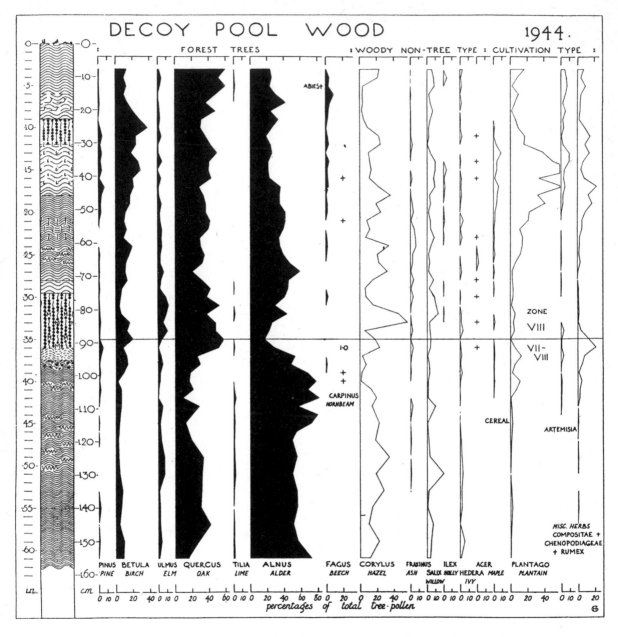

DECOY POOL WOOD   1944.

coincided with building of the Late Bronze Age trackways, the upper with the pre-Roman Iron Age.

Four of the non-tree pollen curves gave clear evidence of prehistoric forest clearance and husbandry, namely those of the large cereal-type grass pollen, of ribwort plantain, of wormwood, and of the aggregate pollen of such weed types as many docks, Compositae or Chenopodiaceae. These four curves were highly correlated with one another and showed two distinct phases of clearance and husbandry, a minor one in the Late Bronze Age and a much larger one in the pre-Roman Iron Age. Each phase seemed to have terminated with the onset of severe flooding, a result likely enough when we reflect that the second clearance phase would correspond, at least roughly, with the occupation of the Glastonbury Lake Village with its abundant cereals and evidence of sheep husbandry. It now seems likely, in the light of events since the diagram was first described, that most of the changes in the shrub and tree-pollen frequencies were also associated with the human exploitation of the region. At the lower flooding horizon the sudden decrease in alder was likely to have followed the enormous demand on the local fen-woods for extensive trackways; this allowed pollen of upland trees to be correspondingly better represented. After this first onset of clearance, opening of the woodlands seems to have been accompanied by spread of birch, ash, holly, maple and beech: only the last of these

◀

Fig. 67. Pollen diagram and stratigraphy at Decoy Pool Wood, Shapwick Heath. This was the first site on the Levels at which a continuous history of agricultural activity had been traced by suitable pollen indicators. In particular ribwort plantain (*Plantago lanceolata*) indicated pasture, and wormwood (*Artemisia*), docks (*Rumex*) and other herbs indicated arable cultivation along with the direct identification of cereal pollen. The correlations which follow are set out in the text as in Fig. 68.

Fig. 68. General correlation schema for events registered in the peat bogs of the Somerset Levels (1963). This indicates that the Neolithic trackways arc of rather varying date, probably because the spread of fen-woods and the origination of raised bog occurred at different times, depending on local conditions. The varying intensity of agriculture is indicated from the Neolithic onward and it is shewn that two of the horizons of wetness in the Levels correspond with Swedish bog regeneration surfaces.

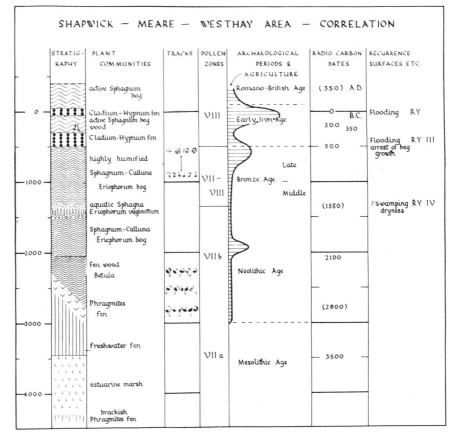

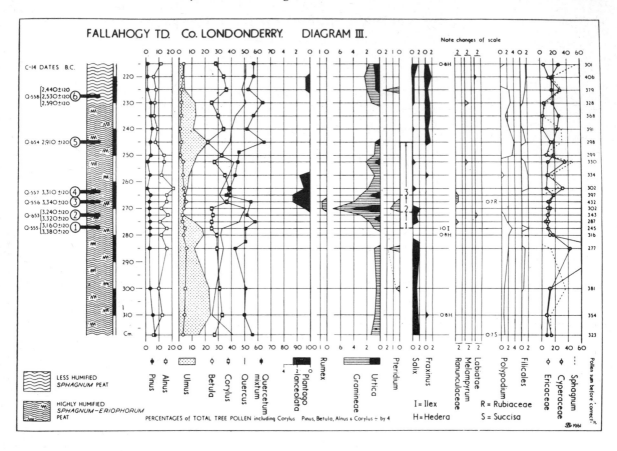

FALLAHOGY TD. Co. LONDONDERRY. DIAGRAM III.

seems in any way a surprising response. It fell to Dr Judith Turner, one of my research students and later, research fellow of Newnham College, Cambridge, to shew later on that the sharp diminution of linden pollen at the first episode was almost certainly also a consequence of clearance here as in other parts of Britain.

The study of the spread of forest clearance and early agriculture by pollen analysis now became widespread, especially throughout Europe, and it was elaborated in many ways. In particular Troels-Smith very forcefully attributed the distinct and large decrease in elm pollen (the elm decline) shewn by so many European pollen diagrams, to selective gathering for fodder of elm foliage for cattle penned up in stalls. He rightly drew attention to this leaf-fodder economy as one persisting in India and even in Europe, and published pollen diagrams that shewed the elm decline preceding the expansion of indicators of cultivation, as if a phase of keeping foliage-fed cattle had gone before the cultivation *Landnam* phase described by Iversen. It was a suggestion compatible

◀

Fig. 69. Joint application of pollen analysis and of radiocarbon dating to the recognition and absolute dating of two phases of woodland clearance and agriculture of Neolithic date in a raised bog in Northern Ireland. The dates are given in radiocarbon years b.c. (as it would now be written). The sudden decrease in frequency of elm pollen shews forest clearance, the considerable amounts of weed pollen (plantain, grass and nettle) shew the cultivation phase, and revertence to previous values shews regeneration of the woodland structure. The whole cycle was over in a few hundred years. The earliness of the first episode (*c.* 3300 b.c.) was remarkable. (From Smith and Willis, 1961/2.)

with the fact that cattle are naturally browsing rather than grazing animals. Similar effects were recognised in 1963 by Dr Frank Oldfield in pollen diagrams from Thrang Moss in Lancashire. It was now evident that the elm decline, so very pronounced in all British pollen diagrams, especially those of the west, might not be primarily a consequence of climatic change as had been initially supposed, but might reflect selective destruction of the elm either by preferential gathering or by clearance of those specially fertile soils where elms grew most abundantly. In approaching this and many other problems of the nature of the forest clearances and spread of agriculture it proved that the peat deposits of the raised bogs offered special advantages, by no means only in the Somerset Levels where it was evident that virtually every pollen diagram made was influenced in one way or another by human influence upon the surrounding countryside. As we have seen, we had now solid proof that hazel coppice was being maintained during Neolithic time, and it is tempting to speculate that this came into being as a consequence of the practice of leaf-fodder collection. Certainly I had seen in the Danube flood-plain in 1930 communities of hazel–hornbeam scrub maintained for leaf-fodder and looking very like British hazel coppice some 20 ft (6 m) high. Dare one speculate that the hazel scrub of western Ireland may also have had some such treatment in the past? Once established, no doubt the practice had such virtues in providing small timber for general use about the countryside and farm, that it naturally persisted.

At Fallahogy in County Londonderry Dr Alan Smith of the University of Belfast* had shewn how a large raised bog near the River Bann yielded a remarkably informative record of a *Landnam* episode, once again at the Atlantic/Sub-boreal transition, differing however from the classic Danish sites in that evidence for cereal cultivation was absent. He had employed counts of no less than 1000 tree-pollen grains per sample, and as in Iversen's sites, three stages were clearly displayed: (*a*) forest clearance with decline of *Ulmus* and rise in grass pollen frequency, (*b*) farming, characterised by sharp increase in the pollen of ribwort plantain and such weeds of arable cultivation as nettle (*Urtica* sp.) and sorrel dock (*Rumex* sp.), (*c*) woodland regeneration with rise of hazel pollen to maximum frequency followed by that of elm, whilst weed and grass pollen decline. Ultimately, with decline of the hazel component, the original woodland composition had been restored.

In an attempt to attach objective estimates of duration to these

---

* Dr Smith was formerly of the Sub-department of Quaternary Research in Cambridge and presently Professor of Botany in Cardiff.

ecological phases, as well as to determine the absolute age of the elm decline, Alan Smith in 1961 collaborated with Eric Willis of the Cambridge carbon dating laboratory. A complete duplicate peat monolith was retrieved from the Fallahogy site, and fresh pollen counts ensured that all the horizons previously described could be precisely recognised, thus allowing the extraction of thin slices of peat for carbon dating at each crucial level. The material was highly humified *Sphagnum–Eriophorum* peat almost ideally suitable for carbon dating: the results are shown in Table 1, the historical sequence running from below upwards.

Table 1. *Neolithic forest clearance at Fallahogy*

Sample no.	Pollen-analytic evidence	Radiocarbon age b.c.
Q-654	End of stage 3: forest regeneration complete	2910 ± 120
Q-557	Beginning of forest regeneration: hazel pollen maximum	3310 ± 120
Q-556	Height of farming activity: plantain frequency maximal (stage 2)	3340 ± 120
Q-653	Near end of clearance (stage 1): elm decline almost concluded	3240 ± 120 3320 ± 120
Q-555	Commencement of clearance (stage 1): beginning of elm decline	3160 ± 120 3380 ± 120

At a substantially higher level in the column an episode of renewed clearance and agricultural activity was proved by a thrice-dated sample to have a radiocarbon age close to 2500 b.c.

The four dates for the elm decline thus provided gave extremely strong indication that this happened around 3000 b.c., and indeed somewhat earlier. Both in Ireland and elsewhere the elm decline had been firmly associated with the Neolithic culture, and that culture had independently received similar radiocarbon dates. It was part of a swiftly accumulating stack of evidence for this considerable age for the onset of the Neolithic period in Britain, a substantial and indeed disconcerting modification of the hitherto accepted view that it should be placed about 1800 B.C.

It is probable on statistical grounds that the uppermost of the dated samples is between 200 and 350 years younger than those which lie below it, and it therefore seems likely that the forest fully healed itself by regrowth after the clearance within a span entirely acceptable in ecological terms. We can only estimate the length of the two earlier stages of clearance and farming from the fact that the thickness of peat they embrace is about one-quarter that of the whole cycle, possibly only a decade or two, improbably more than a century or so. These determinations *in toto* are entirely plausible in an ecological sense, and especially the longer time required for the final forest regeneration stage with the regrowth of young trees, largely from seed.

We have to envisage the whole landscape of lowland Britain before invasion by the Neolithic farmers as mantled with continuous green forest in which the lakes and bogs appeared as infrequent gaps with fens around them and alongside the rivers. In both the lakes and bogs the continuous deposit of organic material caught and preserved the air-borne pollen, so that both have served as archives incorporating the history of the surrounding forest before and through the onset of its clearance. In pollen-analytic terms the lakes have the advantage that little or no vegetation grows locally upon them: only lake-side fen-woods can confuse the interpretation of the pollen analyses, whereas bogs sometimes carry local growth of birch and pine as well as having fen-woods in their marginal lagg areas. The trees apart, the surface vegetation of the raised bogs is so extremely characteristic that its pollen is unlikely to be confused with that of lowland forest, though it may well resemble that of upland mountain and heath. It might be urged that into the lakes there arrives a certain proportion of water-borne pollen, that currents and wave action may disturb and redeposit the soft bottom sediments and that detritus-feeding animals to some extent will mix together the accumulating muds. All the same, the lakes yield such consistent pollen diagrams that these effects cannot be very serious.

When one seeks to combine pollen analyses with radiocarbon dating, the raised bogs seem to offer large advantages. Any hard-water lake is liable to the risk that its organic muds carry a proportion of inactive, ancient carbon carried in by the drainage water as bicarbonate and fixed by the photosynthesis of submerged plants, so that it enters the chain of feeding and decay that finally ends in the lake muds themselves. Mixing within the lake muds by erosion and drifting is also a serious potential source of error leading to unsuspected gaps and repetitions in the sequences. The apparent handicap of taking clean samples in a substantial depth of water has now been largely overcome by elegant samplers operated from rafts or other vessels, or even through holes cut in the surface ice of a severe winter.

By contrast, raised bogs are entirely made of organic material whose carbon has been directly fixed from the atmosphere: there is no hard-water error. Although there is some transfer of younger material downwards through the penetration of growing roots and some down-leaching of humic material in solution, these are relatively minor effects and can to some degree be offset by field and laboratory treatment. Moreover the rapid growth rate of most peat bogs allows a very useful vertical separation of the successive pollen and carbon dating samples, and we have already set down abundant evidence of the way in which the bog-growth will incorporate and isolate in its successive layers climatic, archaeological and geological, as well as vegetational evidence.

It was reasons of this kind that ensured that in the further pursuit of joint radiocarbon and pollen-analytic studies, raised bogs should, to a considerable extent, be preferred sites.

This was particularly so with the work of Dr Judith Turner. Her re-examination of the Shapwick Heath pollen sequence across the elm–linden decline, dated about 1900 to 2000 b.c., showed such decisive correspondence with increases in the frequency of pollen of herbs and of bracken spores, along with spread of the light-demanding ash, that it seemed most probable that the decline of both elm and linden, in this instance at least, was anthropogenic and not climatically caused. Judy now turned to reconsider the pollen evidence at those other raised bogs where the *Tilia* (linden) decline had already been demonstrated, and was able to prove different dates for the phenomenon at the four sites, with the bonus that at the Yorkshire site the *Tilia* decline was found at two successive levels (Table 2). The close intervals of the samples and the large pollen counts enabled Judy to estimate at the Whixall site the time taken for forest clearance and establishment of open communities of grassland rich in ribwort plaintain. The approximate figure of 150 years for such a change is of the same order as that at Fallahogy.

Table 2. *Age of the* Tilia *decline*

Site	Sample no.	Radiocarbon age b.c.
Shapwick Heath	Q-644	2015 ± 115
(Somerset)	Q-645	1920 ± 115
Holme Fen (Hunts.)	Q-403	1440 ± 120
	Q-404	1445 ± 120
Whixall Moss (Shrops.)	Q-467	1277 ± 115
Thorne Waste (Yorks.)	Q-481	1210 ± 115
First decline	Q-482	971 ± 115
Second decline	Q-479	368 ± 110

The overall effect of this study was of course to cast grave doubt on the presumed climatic cause of both the elm and linden declines of whatever date, but we have to bear in mind that where the elm decline has been roughly synchronous over great areas, as throughout most of the British Isles around 3000 B.C., and has been accompanied by other vegetational indications of climatic shift less susceptible to the *Landnam* explanation, it may be wise to ask whether an underlying major climatic shift may not indirectly have caused westward migration of farmer folk at the given time, so that there remains a double and complex causative mechanism. It has long been understood by both historians and archaeologists how a global shift of climate towards greater severity was the presumed cause of the successive waves of invasion of the West by the hard-pressed nomadic inhabitants of central Asia. Such movements are inevitably transmitted by migrations in turn outwards from the invaded territory and the British Isles have not been immune from them, nor indeed are they now.

The pioneer German palynologist who had first made use of cereal pollen identifications, Franz Overbeck, had already pointed out how his own pollen diagrams appeared to reflect vegetational changes accompanying such historic events as the foundation of local monasteries, the Black Death (between A.D. 1350 and 1500) and the Thirty Years War (A.D. 1632–1643). It was in an attempt to find whether comparable effects might be recognisable in Britain that Judy now returned to the raised bogs at Tregaron where we had already a general knowledge of the situation. By choosing a site on the south-eastern bog, not far from the margin, it could be expected that one would find the pollen record registering separately any local clearances on the contiguous hillsides: a site far out on the bog complex must, by contrast, have caught a general pollen-rain in which many clearance and occupation episodes would have been jointly shewn, the effects inevitably combined and confused. From the south-eastern bog site a clean column of peat was cut and transported to the laboratory: from it pollen samples were taken at extremely short intervals (in some sections as little as a quarter of an inch, 6 mm) and subsequently, at the critical horizons, thin peat slices were used for radiocarbon dating.

As work proceeded it became evident that one could make a good assessment of the origin of various groups of pollen, the components of which always seemed to behave similarly. These were as follows:

A *Local pollen* (bog surface) e.g. *Calluna*, sedges and sometimes bracken
B *Pollen from upland* Bi : undisturbed woodland
                        Bii: clearance indicators
                             α: pasture
                             β: arable cultivation

The clearance indicators taken to be indicative of pasturage were particularly the grasses, ribwort plantain and the docks. Those indicative of arable economy included very miscellaneous herbs of the families Cruciferae, Leguminosae, Compositae and Chenopodiaceae, with such characteristic elements as *Artemisia* (wormwood) and *Centaurea cyanus* (cornflower).

In Cardiganshire (Dyfed) at the present day about 80 per cent of the total acreage is under grassland as against 20 per cent arable, whereas in the counties of Norfolk and Suffolk these frequencies are reversed. Judy now found strong confirmation in her classification of the herbaceous pollen types by analysing, from such materials as living moss-clumps, the present-day pollen catch at sites representative of these two respective regions. It was remarkably clear how today's pollen-rain in Cardiganshire had some 80 per cent of the pasture indicators whilst in the eastern counties it was the arable indicators giving such values. This result gave confidence to the provisional reconstruction now derived from the Tregaron pollen diagrams, as set out in Fig. 70. The correspondence with what is known of local history is very convincing.

In the prehistoric period it seems that there were passing episodes of temporary and local clearance from which the woodlands could recover and that these extended into the Iron Age when a short period of extensive disforestation introduced the system of pastoral farming

Fig. 70. Diagram to shew the relation between the conclusions based upon pollen analysis and available historical data for the vicinity of Tregaron Bog. The depth of the peat samples is shewn on the left: the radiocarbon dates were given by the Cambridge laboratory and have standard errors of 90 to 110 years. The correspondence between the two sources of information on former agricultural practice is remarkable. (After J. Turner.)

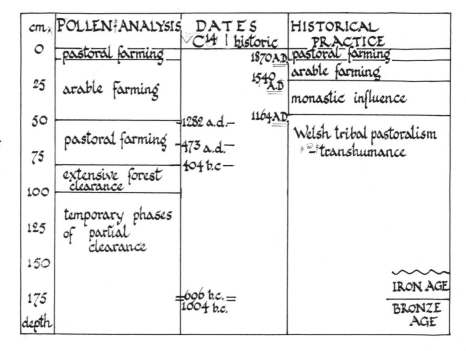

which is familiar from early historical records. About 1164 there was founded nearby the famous monastery of Strata Florida, so strongly associated with the legend of the Holy Grail, and it is at this time that there seems to have been established a substantial shift to arable farming, probably due to the associated granges rather than the mother church itself. Finally the uppermost part of the record represents the recent historic change in emphasis towards pastoralism.

Similar exercises extended to Scottish bog sites in Bloak Moss (Ayrshire) and Flanders Moss (Perthshire), established that there again the age and character of the forest clearances could be recognised in the dated pollen diagrams. In seeking to recognise the consequences of human modification of the natural forest vegetation, sometimes one is concerned with economic usages that hardly modified forest composition at all, perhaps preventing pollen production in a single tree-species, and merely making use of the forest products, sometimes a system of woodland management such as coppicing and sometimes a form of shifting cultivation not permanently altering the woodland composition. On the other hand there were the systems of usage in which woodland was removed and replaced by communities of pastoral or arable agriculture, when the soil and site were deflected altogether from a woodland economy. There is great potential in the pollen diagrams for recovery of the story of such past systems of husbandry and exploitation: it goes without saying that great caution and experience are demanded in drawing final conclusions where the situations and evidence are both so complex.

13

Pollen zones and sea-level changes absolutely dated

Pollen zones

Already by 1935 the many pollen diagrams in existence for all Europe showed a remarkably consistent pattern of forest history which was most easily interpreted as a northward migration of broad vegetational zones under the compulsion of the progressive improvement of climate after the ice-age, up to a northern limit from which there had been some withdrawal to the present natural limits. Corresponding shifts of altitudinal range had accompanied these changes, the forest belts ascending the mountain flanks as the climate warmed and descending as it cooled. There were also movements towards the Atlantic or towards the heart of the continent.

Following von Post's lead there were set up for various parts of Europe numerical pollen zones such that a zone of given number was equivalent, although not vegetationally identical, over a wide territory. It ought to be clearly said that it was the intention from the outset to make these zones broadly synchronous: this was the prevalent usage of the concept of a zone in geology, i.e. a belt or ring of some spatial extent whose stratigraphic position, and hence age, was conveyed by its fossil content. The same intention is disclosed by the way von Post linked his zonation to the threefold division of Post-glacial history over large parts of the world into periods of (i) increasing warmth, (ii) maximal warmth, (iii) revertence to cooler conditions (cf. fig. 3).

Thus from the outset there arose a numerical zone system for the Late-glacial and Post-glacial time, whose purpose was largely to supply a chronology for a period otherwise without one, and evidence of stratigraphy, lithology and biogeology was often adduced to support the reference of a particular horizon to a given zone. The situation was not basically affected by the chosen numerical zones being those favoured by the German and Danish research workers rather than that of von Post which kept zone I for the uppermost end of the sequence rather than the other way round.

When, in 1940, I attempted an analysis of all the pollen-analytic evidence then available for England and Wales it was a zonation of this kind that was suggested. I made it correspond as far as possible with the zones of the neighbouring continental mainland, and when Knud Jessen published the synthesis of his work in Ireland (1949), he took trouble to retain that correspondence. The proposed zone system was widely adopted in Britain and proved of great value as at least a quasi-chronology of the Post-glacial period. It has, however, become clear that the natural rates of migration of forest trees are sometimes slow in relation to the climatic amelioration and that the quick expansion of a tree such as alder at the opening of a given pollen zone might be read either as a direct response to climatic change towards warmth or wetness, or alternatively to its first arrival *en masse* in the region. That is to say that the sequence of expansion of the forest elements might be earlier in the south of the country merely because they arrived there soonest. If this were so the zone boundaries were not synchronous but of different ages (metachronous).

The successful development of radiocarbon dating put into the hands of the Cambridge Sub-department of Quaternary Research a new and most powerful tool for determining the *absolute* age of the major and minor pollen zones and so eventually of attacking the fundamental problem of the country-wide synchroneity of pollen zones. Donald Walker, later Professor at the Department of Pacific Studies of the National University of Australia, had in 1950, in the course of studies on the peat deposits of the Solway area, discovered a relict raised bog highly suitable to our purpose, Scaleby Moss, lying about 5 miles (8 km) north of Carlisle in a depression in Boulder Clay of the last glaciation. He had secured from it a detailed pollen diagram that beautifully displayed the whole sequence of regional vegetational history and that was readily zoned (Fig. 71). In 1955, he, Eric Willis and I decided to exploit this site further. Most of the original bog surface had been cut away, leaving an irregular top colonised by heather, cotton-grass and *Sphagnum*, with scattered birch and pine, but at one locality there remained a plateau of uncut peat extending to the original bog surface, and it was next to this that Donald Walker and I began operations. From the cleaned face of the plateau of younger peat we secured in steel ammunition boxes, a sequence of vertical columns of undisturbed peat each measuring 45 × 18 × 17 cm. To extend this sequence downwards we now dug a pit some 150 cm square and 300 cm deep. It called for the tidiest of peat-cutting techniques and some sustained heavy labour, but finally presented us with a clean lateral face into which the steel sampling boxes could readily be pressed home.

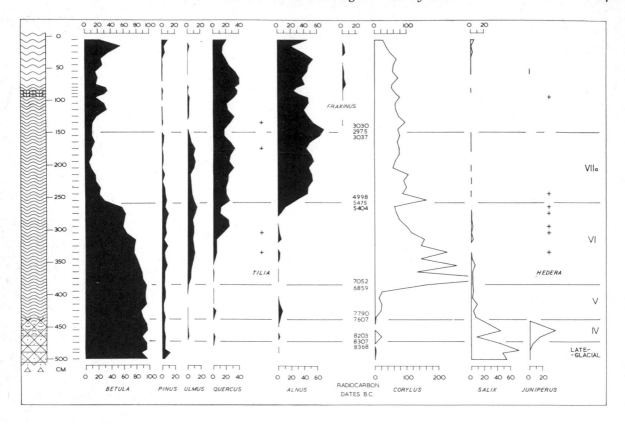

Fig. 71. Scaleby Moss, Cumberland. The first long British pollen diagram to have the pollen zone boundaries dated by radiocarbon assay. Contiguous samples agree well and the results are self-consistent in sequence. The radiocarbon ages would now be designated as b.c., since they are based on uncorrected radiocarbon years. (From Godwin, Walker and Willis, 1957.)

We finally had, including miscellaneous samples, nine or so very heavy steel boxes, our spades and peat-boring gear to be removed from the bog site to the firm road over a mile away. I recall how we brought into service a 'dolly' used for carrying milk-churns, and how we sweatily urged this small-wheeled vehicle over the hummocky bog surface under the baleful red glare of the restless bull who was temporarily in possession of the intervening rough pasture. The kind north country ministrations of the farmer's household and of the friendly curator of the Carlisle Museum set us up for the return journey with our carload of booty to the Cambridge laboratory.

Once in the laboratory, closely spaced pollen samples from the monoliths of peat enabled us to identify the precise levels of all the pollen zone boundaries in the original diagram. At each such level we cut one or more horizontal slices of peat, 2 cm in thickness, and these were processed for radiocarbon assay, in the first instance by oven-drying them at 100 °C and storing in a sealed polythene bag. The final results were sixteen radiocarbon dates extending between about 3000 b.c. and 9000 b.c., and they shewed a gratifying internal consistency. In the first place the determined ages corresponded fully to the depositional order. Secondly, wherever there were contiguous samples the radiocarbon dates were in very close agreement, and finally the results conformed well with other known indices of age including other early radiocarbon age determinations. We now had, for the first time in a British site, an objectively dated long pollen zonation. Its main zone boundaries were six, as shown in Table 3.

Table 3. *Pollen zone boundaries at Scaleby Moss*

	Approx. radiocarbon age b.c.
(*a*) The VIIa/VIIb horizon – the elm decline – the Atlantic/Sub-boreal boundary	3000
(*b*) The VI/VIIa horizon – the 'alder rise' – the Boreal/Atlantic transition	5300
(*c*) The V/VI horizon	6950
(*d*) The IV/V horizon: Pre-boreal/Boreal transition	7700
(*e*) The III/IV horizon: Late-glacial/Post-glacial transition (opening of the Flandrian)	8300
(*f*) The II/III horizon: end of mild interstadial Allerød	8800

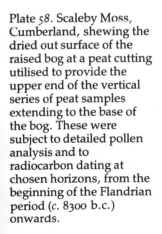

Plate 58. Scaleby Moss, Cumberland, shewing the dried out surface of the raised bog at a peat cutting utilised to provide the upper end of the vertical series of peat samples extending to the base of the bog. These were subject to detailed pollen analysis and to radiocarbon dating at chosen horizons, from the beginning of the Flandrian period (*c.* 8300 b.c.) onwards.

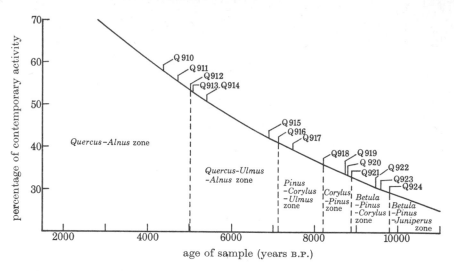

Fig. 72. Red Moss, Lancashire. Upon the natural decay curve for radiocarbon have been set the determinations of the critical samples chosen to represent the major pollen zone boundaries. The zones are here given descriptive names: the ages are in radiocarbon years before the present (1970). The results broadly confirm those obtained at Scaleby.

The Dutch carbon dating laboratory at Groningen had already obtained similar dates from Dutch and German sites for the elm decline horizon and for the Boreal/Atlantic transition, but we were particularly pleased to note that the zone III/IV boundary had been dated in Copenhagen at about 8350 b.c., whilst quite independently this horizon had been given a date of 8100 b.c. on the basis of the Scandinavian annual laminations of ice-front lakes, the so-called varve chronology.

The Scaleby Moss results thus appeared at once to validate the radiocarbon dating process and the suitability of the raised bog as source of material appropriate to it and to pollen analysis alike. The way was open to move into the next stage of determining whether the zone boundaries are of similar age over the country as a whole or whether they are 'sloping', i.e. metachronous. In point of fact, although small sequences were often dated, it was not until 1970 that another raised bog comparable with Scaleby Moss was investigated by three of our successors in the Sub-department. This was the Red Moss, transected by motorway excavations where the M61 crosses north Lancashire. The sixteen dates obtained here ranged between about 2420 b.c. and 7850 b.c., and although they attained considerably smaller probable errors, the results in general were in agreement with those at Scaleby.

Both of the bog sites sufficed to yield an answer to the questions that were most frequently presented to one in the early days of concern with peat and peat bogs, to wit 'how fast does the peat grow?' and 'how long does it take the bog to form?' At Scaleby Moss the mean rate of growth of the peat was just over 6.0 cm per century, but at Red Moss the figure for the older peat was 3.3 cm per century and for the younger 4.2 cm. Of course the measured thicknesses have been affected in each site by the history of humification, compression, cutting and drainage to an unknown extent.

It is now possible to recognise retrospectively how important to Quaternary research was the stage represented by the Scaleby Moss exercise. It was then, 1955, that one was finally assured of the broad effectiveness of the pollen zone system and of the chronological frameworks that had been evolved to sub-divide the Flandrian period, frameworks such as we described earlier for the Somerset Levels, based on integration of evidence from the most diverse sources. It was inevitable that research workers from this time forward, always able to refer the crucial questions of age to a few radiocarbon samples, should shape their research problems in quite a new way; they were now naturally able to move about the Flandrian period, and for that matter the preceding Weichselian (Glacial), with altogether fresh assurance.

Registration of sea-level

As the oceans of the world filled up again with the water restored to them by the melting of the great ice-sheets of the last Ice Age, shore-lines moved relentlessly upwards and estuaries were extended by the invasion of sea-water. We have seen the effects of this process in our account of the geology of the Somerset Levels. At the height of the

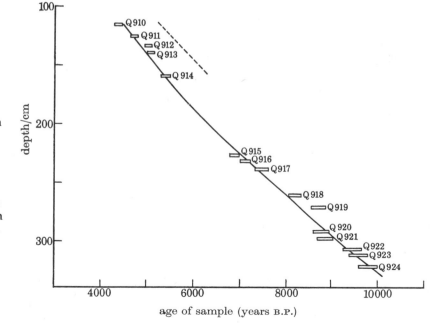

Fig. 73. Red Moss, Lancashire. The radiocarbon dating of a detailed pollen series from the raised bog allows one to plot the rate of peat accumulation throughout growth of the bog and over a span of more than 5000 years. The overall rate is about 3.8 cm per (radiocarbon) century, but possibly two slopes can be distinguished, shewing more rapid growth in the upper layers. (From Hibbert, Switsur and West 1971.)

transgression, behind coastal bars of sand and gravel the lowest reaches of the rivers were often converted to shallow lagoons in which the tides, day by day, laid down a deposit of silts and clays ultimately to form salt-marshes or brackish-water reed-swamp just above mean sea-level. It was indeed in the flatness of these large areas of tidal mud that the 'Levels' had their origin, and the marine influence once altered to regression, it was the extreme levelness which then led to those ecological developments already described. The lagoons became firstly brackish-water *Phragmites* swamp, then, as salinity diminished, fresh-water sedge-fen, in turn colonised by fen-woods of sallow, alder and birch, that gave foothold for the development of raised bogs. These mires, whose main constituent plants are intolerant of even low concentrations of bases, indicate full freedom from incursion of calcareous water. How much more then are they indicative of the absence of sea-water flooding, although often persisting, as we have indicated in Somerset, Borth Bog and Swansea Bay, quite close to the present-day coast. If, by the vagary of coastal erosion, sections or drainage cuts like the Huntspill Cut or the straightened River Leri at Borth Bog, are provided which display the stratigraphic sequence, they insistently challenge the geologist to decipher from them the past history of land- and sea-level changes. As on the foreshore at Ynyslas beside Borth Bog,

Plate 59. The submerged forest exposed on the foreshore at Ynyslas on Cardigan Bay. Various large stools, chiefly of oak, are rooted in a thick peat bed, now partly obscured by shifting sand of the modern beach. This peat bed is continuous with the basal layers of Borth Bog that lies a short distance inland behind the present-day shingle ridge of the beach (see Figs. 19–21).

Plate 60. Borth, looking west across the great expanse of raised bog to the coastal dunes that alone separate it from the shore of Cardigan Bay. Tidal estuary of the Dovey on far right with rocky outcrops in middle distance. (1959)

coastal erosion may indeed present evidence in the shape of a 'submerged forest' with sub-fossil tree-stumps still in position of growth upon a peaty substrate. As we have already shewn, this turns out in fact to be merely part of the fen-wood vegetational succession that formed one stage of the development of the raised bog, of which it is now only an isolated remnant. 'Noah's forests' have naturally long been part of folk evidence for past inundation to those around Cardigan Bay and no doubt contributed to belief in 'Cantref y Gwaelod', the lost land of Wales, a legend that might well have its foundation in transmitted recollection of the actual marine incursions.

As long ago as 1910 the pattern of raised bogs sitting upon the flat surface of estuarine clays had been recognised in the lowlands beside the River Kent estuary at the head of Morecambe Bay in north Lancashire. These mosses, formerly of great extent, have been much affected by drainage, peat cutting and reclamation, but their essential character remains evident and here and there stratigraphic sections can be recovered from them, whilst in some instances they still harbour valuable populations of the plants and insects that flourished there when the bogs were still growing. When, about 1952, Alan Smith undertook reinvestigation of three of these bogs, Helsington Moss, Foulshaw Moss and Nichols Moss, using our new methods and drawing on recent research, he was able quickly to shew that in stratigraphic development they followed just the same pattern as that which had now become clear in the Somerset Levels and at Borth. He determined that the level of the surface of the estuarine clay in each of the three was between +12.4 and +15.5 ft (+3.8 and +4.7 m) OD and his thorough pollen analyses made it entirely clear that all three bogs had originated at a closely similar time, that is to say early in the English pollen zone VIIa and accordingly within the first part of the Atlantic period.

A still more precise definition of this important geological horizon, the top of the marine transgression, was provided in the neighbouring Winster Valley where a solid rock bar was found to exist, cutting off an upper basin, still occupied by Helton Tarn, from Nichols Moss in the seaward part of the valley. At the maximum of the ocean rise, salt water just overflowed the bar to lay down in the tarn a thin deposit of clay, whose age was proved pollen-analytically to agree exactly with that of the initiation of the three coastal raised bogs. The rock-sill, proved by levelling to be at 16.6 ft (5.0 m) OD, provided an extremely close estimate of the contemporary ocean-level and reinforces the figures already obtained.

The evidence of the Lancashire bogs had now fallen into accord with that accumulating elsewhere that the great eustatic rise of ocean-level was terminated during the Atlantic period, with the final fast wasting of the North American ice-sheet. The relationship observed between land- and sea-level, however, required explanation in terms of another feature of the waning glaciation. This is the so-called isostatic adjustment of the earth's crust to the local burden of the ice-sheets upon it. In northern Britain, as in Scandinavia, there is clear geological evidence that in the regions where ice accumulation was greatest, there was the greatest upwarping of the crust after the ice melted away. This is most conspicuously displayed in the obvious 'twenty-five' foot raised beach of Scotland and Northern Ireland, a structure that is said to reach 35 ft (17 m) OD at its highest and descends to present sea-level in northern England, Ireland and the western Scottish Isles. It is generally held that it formed during the time when the eustatic rise in ocean-level was slowing down and for a time kept pace with the isostatic recovery. Subsequently the first component no longer acting, the beach has been uplifted to its present position where it is naturally exploited by local coast roads and caravan sites.

The height which had now been proved for the top of the eustatic rise fitted perfectly into this explanation. In Somerset and South Wales the top of the eustatic rise is close to present sea-level whilst Alan Smith's results show it in north Lancashire at around 4 m, that coastline being within the known area of isostatic uplift. We were soon able to match the radiocarbon dating for Somerset and north Lancashire and to establish that the top of the estuarine clay was roughly 3500 b.c. in both

Plate *61*. The Post-glacial raised beach of the west coast of Scotland. Caravans take advantage of the flat beach behind which the contemporary cliffs can be seen. South of Girvan. (Photograph by N. T. Moar.)

areas. Naturally enough there was no trace in the Lancashire area of relative land-elevation, of the later marine transgression seen in Somerset in the Romano-British period, or that in the upper layers of Borth Bog already mentioned in Chapter 3.

As the raised bogs were proved to have formed upon the estuarine clays of the major eustatic rise in north Lancashire, so they were proved to occur in the submerged estuaries on either side of the Solway Firth. More dramatically still, the great Flanders Moss in Stirlingshire is sitting upon the surface of the so-called Carse Clays that formed when the rise of ocean-level extended sea-water right across the Scottish Lowlands between the Firths of Forth and Clyde. These clays, long famous for the whale skeletons occasionally found in them, are recognised as having formed at the time of the Scottish Post-glacial raised beach. The reinforcement of bog stratigraphy and pollen analysis by radiocarbon dating has given much greater certitude to our reading of the evidence of the coastal raised bogs already apparent in outline in the mid-1950s.

This evidence was always more abundant in the north and west: in the south and east not only was the drier climate less favourable to bog growth, but pressure of demand for fuel has meant almost total destruction of what peat deposits were available. I recall the surprise with which I examined a sample of the peat bed laid bare on the Norfolk coast by the disastrous flooding of 1938 when destruction of the coastal dune system at Horsey and Palling allowed inundation of the upper reaches of the hinterland Broads of the Thurne valley. The sample was a typical well-humified *Sphagnum–Calluna–Eriophorum* peat testifying to the former presence of East Anglian raised bog, however unexpected that then might be. Likewise acidic bog deposits have been recognised only infrequently and at the landward margin of the East Anglian Fenland, and it was not until 1960 that Dr Joyce Lambert and her colleagues were able fully to establish that the Norfolk Broads had been created by the mediaeval digging away of raised bogs that formerly existed in the river valleys. It is not surprising, therefore, that in the southern half of Britain the raised bogs should be only of minor importance as revealing past alterations of sea-level.

One possible exception may, however, be noted. Along the south coast of England the River Arun cuts through the South Downs, and records of geological borings in its valley, as well as the discovery of prehistoric boats in its deposits, gave reason to think that here also there was opportunity to trace Post-glacial sea-level changes. This was confirmed when Roy Clapham and I visited the area in 1940, discovering that the well-known naturalists' haunt, Amberley Wild Brooks, was nothing less than a derelict raised bog still supporting dead *Calluna* bushes upon its recently ploughed surface although no longer supporting the creeping *Oxycoccus* that had caused the naming of the adjacent 'Cranberry Farm'. The bog deposits rested upon the surface of an estuarine clay that had, during a previous marine transgression, filled a lagoon behind the Downs to a height of +3 to +8 ft (1.0 to 2.5 m) OD. The pollen diagram indicated that this transgression ended late in zone VIIb, that is close to the Sub-boreal/Sub-atlantic transition, a conclusion to be confirmed many years later by a radiocarbon date of 670 b.c. ± 110 years. It was evident that here we were concerned with an event much later than the major eustatic rise in ocean-level, and one in age at least coming closer to the Romano-British transgression seen in Somerset and the Fenland. The relative downwards tilt of the southern North Sea basin mentioned already (Chapter 10) has carried deposits that might register the end of the main eustatic rise so far below present sea-level that they must exist only in very deep valleys or in sites out at sea.

14

Climatic registration

Recurrence surfaces

Because the acid bogs are ombrogenic in nature, that is to say dependent upon the precipitation reaching them, they must always be sensitive to changes in rainfall and evaporation. It is not surprising, therefore, that in the successive peat layers of their upward growth they should turn out to have registered former climatic changes. We have learned already from the raised bogs of the West Country that such changes were recognisable in the macroscopic plant residues of the peat, in its degree of humification and in the statistical composition of its pollen content. In particular we have noted that alike at Tregaron, Borth and the various exposures of the Somerset bogs, one finds the twofold division of the peat into a lower humified 'black' peat and an upper fresh and less humified 'white' peat. The division, familiar enough to the peat cutters, is an abrupt one and we have seen that it corresponds with the phenomenon known in Germany as the *Grenzhorizont*. This phenomenon, that we may call a Boundary Horizon, is so widespread not only in the British Isles but throughout north-west Europe, that it is hard to suppose that it is not the response to some common cause, such as widespread climatic change. This indeed is the explanation put forward originally by the pioneer of bog stratigraphy, C. A. Weber, and the cause has generally been attributed to the change, about 500 B.C., from the dry Sub-boreal climatic period to the succeeding wet and cool Sub-atlantic of the Blytt and Serander sequence.

Weber had thought that in those bogs with which he was familiar, peat growth had ceased altogether in the Sub-boreal, so that there was a time-gap in the peat deposition, but this notion has been steadily discarded and is clearly disproved by the radiocarbon dates throughout such bogs as Red Moss. Here the actual dates through the depth of the bog shew no hiatus although quite often the upper *Sphagnum* peat can be shown to have grown faster than the dark peat below the Boundary Horizon (see Fig. 73). Quite often also there is evidence that the growth

Plate 62. A deep raised bog section at Klazinaveen, near Emmen, north Holland. It shews an extremely striking contrast between a lower humified peat and an upper pale peat. At the junction (as by the spade) trees have grown on the surface of the older peat, and clearly the bog became much wetter after this. Another less pronounced, but consistent surface is recognisable in the upper peat itself, where moderately humified peat gives way to very fresh *Sphagnum* peat. One supposes that these stratigraphic changes were climatically induced. (1959)

of the old humified peat terminated in a phase of colonisation by pine and birch, or by *Calluna* and *Eriophorum*, all indices of a pronounced drying of the bog surfaces that made more dramatic their subsequent water-logging and the development of the precursor peat that inaugurated the return of rapid growth of *Sphagnum*-dominated communities.

It is in north-western Europe, and especially in Scandinavia, that investigations have been most thoroughly pursued into the phenomena of bog stratigraphy, and it was from a Swedish scientist, Granlund, that we had the demonstration that recurrence surfaces were not the product of a single period only, but could be consistently found at five periods at about 2300 B.C., 1200 B.C., 500 B.C., A.D. 400 and A.D. 1200. These were referrred to as RY I, II, III, IV and V. Subsequently the number of such recognisable horizons has been suggested for certain areas as nine, but in Britain we no longer have untouched peat bogs adequate to test this view. There has also been the opinion expressed by Granlund that a recurrence surface is the product of a domed bog growing upwards, in a given climate, to a limiting height, where its convexity has so increased the surface drainage that the bog communities cease to be dominated by active *Sphagnum* and enter a *Stillstand* condition. In fact many raised bogs that have a recorded recurrence surface in them are of such great extent that it is impossible to believe that an increased height of 3 or 4 ft (1 to 1.2 m) in the bog centre would have made any noticeable change to surface run-off. It was also easy to observe at Tregaron that all three of the raised bogs there filling the valley shewed similar recurrence structure whereas on the hypothesis of limited height the smallest north-east bog should surely have formed its recurrence horizon long before the great western bog.

If, as seems likely, the history of the past embraces a continuous story of climatic cycles of varying intensity and duration, it seems entirely natural that prolonged maxima of dryness should indeed be directly induced in all the ombrogenic mires, of course most clearly in sites where regional climate, exposure and elevation make them specially susceptible.

A great deal of support for the direct climatic induction of the recurrence surfaces comes from proof of their similarity in age from one bog to the next, and broadly speaking the radiocarbon dating (our safest reference) confirms this. Likewise one would expect a given recurrence surface to be of the same age across the whole extent of a given bog, but on this issue the evidence is inconclusive, partly because the stratigraphy tends to be diffuse towards the bog margins, partly because pollen analysis is by itself too insensitive and equivocal a test, and partly because exact radiocarbon dating is still not thoroughly applied to the problem.

Although in general the blanket bogs yield less evidence of climatic change in their stratigraphy, over considerable areas they display a similar twofold division into a dark, humified lower peat and an upper

Plate 63. Extensive erosion of blanket bog at Holme Moss, 1700 ft (520 m) in the southern Pennines. The peat shews a strong division into an upper less humified and a lower more humified layer: this may represent the major recurrence surface widely seen in raised bogs. The wet floor of the channel at the mineral surface is reinvaded, chiefly by the many-headed cotton-grass (*Eriophorum angustifolium*).

paler and fresher peat, and this has been equated with the most conspicuous of the recurrence surfaces of the raised bogs, RY III. It was not however until Dr Verona Conway, working from Sheffield University, had undertaken detailed examination of the morphology, ecology and stratigraphy of the southern Pennine blanket bogs and had applied pollen-analytic methods to their study, that they could be seen also to have registered repeated climatic shifts similar to those that had affected the raised bogs.

It is interesting to reflect that just as the great anticline of the Pennines (with the subsequent removal of all the top of the uplift) was the geological structure that determined the presence of the coalfields of Lancashire and Yorkshire upon its western and eastern flanks, thus creating the setting for the intense industrial revolution in these areas, so the uplift, especially by exposing large flat stretches of the sterile Millstone Grit at high elevation, provided climatic and edaphic conditions inducing ombrogenic bog formation. When the great conurbations east and west of the Pennines in due time gave birth to their own independent universities, it was inevitable that scientists from them should be attracted to study the great bog formations upon their doorsteps. Thus the universities of Durham, Leeds, Sheffield and Manchester in particular have made highly rewarding research into the Pennine blanket bogs, the more readily as they built upon the support and expertise of the great amateur naturalist tradition of northern Britain, together with the important journals of the regional scientific societies. Verona Conway's findings were published as early as 1947 and 1948 and were accordingly without radiocarbon dating support. Since this has been available some of the most important research on the Pennine blanket peat has come from Dr John Tallis of the University of Manchester, some of whose conclusions are reflected in the next few pages that briefly consider the climatic implications of the initiation of blanket bog, of fossil-tree horizons and of the erosional phenomena now so prevalent in our upland mires.

Blanket bog initiation

Since blanket bogs are limited to regions which have more than a certain precipitation, about 50 in (1250 mm) per annum in our latitudes, or more exactly, a given precipitation/evaporation ratio, it may readily be supposed that one factor necessary for their growth to begin must be adequate climatic wetness, presumably the initiation following climatic shift in that direction. Thus considerable interest has always been attached to learning the date of the base of the blanket bogs, by whatever means deduced. We have already mentioned the forest layer below blanket bog in western Ireland, and besides this there was much else to suggest that this landscape had not always been hidden below its green and black carpet of peat bog. I recall for instance being taken by Frank Mitchell to see the considerable depth of peat that had formed over the flanks of the ancient passage-grave tumulus since it was built on the summit of Carrowkeel in County Sligo, at an altitude of 250 m, possibly 4500 years ago. A similarly dramatic instance of dating the growth of the blanket bog was made by Frank much later at Belderg, County Mayo, where dark criss-crossing grooves scratched into the sub-soil were recognised as plough-marks of the Bronze Age occupants of the site 3200 radiocarbon years ago. This remarkable evidence has been preserved by subsequent deep burial of this site of settlement of a local group of men who probably were essentially copper miners.

At the time when my own interest was first directed to the acid mires, it was sharply encouraged by a visit to the Cambridge Botany Club of the wise and homely figure of Dr T. W. Woodhead, curator of the Tolson Memorial Museum at Huddersfield. He was a formidable pioneer ecologist, deeply informed on all aspects of the Pennine moors and an enthusiastic student of the deep peats covering that inhospitable land-scape. I remember he drew our attention to the paved Roman road, rutted with vehicular traffic, that went westward across the Pennines from Lancashire: it was now observable at Blackstone Edge, where it rested upon peat and was also covered by its later growth.

This evidence carried back the origin of the blanket peat at least 2000 years, but how much more we were unable to say. Later, in his presidential address to the British Ecological Society (1929), Woodhead reported an occasional discovery of a Bronze Age artefact from the lower part of the peat, a bit of bronze and a barbed-and-tanged flint arrow-head. It was tantalising that in fact there was a great abundance of flint and chert artefacts found throughout the peat-clad Pennines, including a large proportion of the finely worked microliths then referred to as Tardenoisian, part of the broader Mesolithic culture that preceded the Neolithic. Unfortunately, frequent as such remains have proved to be, they then seemed invariably associated with the underlying mineral soil so that they afforded for the peat only a '*terminus post quem*', and a very broad estimate at that. A similar situation was found by Elgee to prevail in the Cleveland Hills.

It will of course be guessed that with the advent of pollen analysis hopes were high of solving the age of the Pennine peats, and Woodhead reproduced early diagrams made by Erdtman from Warcock Hill, one of

Plate *64*. Site of Mesolithic occupation at Stump Cross, Yorkshire, in the southern Pennines. The area is now covered with blanket bog but where this large erratic boulder was perched there was abundant evidence of an occupation site associated with a local pool that yielded a vertical series of samples for pollen analysis and a recognisable occupation horizon (see Figs. 74 and 75).

the most instructive archaeological sites. It does not go far, however, beyond demonstrating that the peat at that point was of Post-boreal age. It was not until 1955 that the opportunity presented itself to recover the Pennine microlithic industry from an undisturbed peat sequence. This was at Stump Cross, near Grassington, Yorkshire, at an altitude of 1200 ft (365 m) OD, where a local concentration of flint and chert tools had been found. Here it happened that a large glacial erratic of Millstone Grit had been naturally split vertically to create in the adjacent surface a deep pool that had gradually filled with organic muds laterally continuous in their upper layers with the surrounding acidic peat. Pollen analysis of these muds by Donald Walker showed that they had accumulated from the end of the English pollen zone V to zone VIIb. Mesolithic flint flakes of the typical Pennine culture were recovered directly from the freshly exposed peat sequence at levels corresponding with the period between the zone VI–VIIa transition and the middle of VIIa, that is to say within the opening of the Atlantic climatic period, a finding in close agreement with the increasing attribution of the Mesolithic cultures in lowland Britain also to Boreal or early Atlantic time. Here at this site, so fortunately exploited, it was possible to recover from the artefact level, sufficient organic mud and charred wood for a radiocarbon assay that turned out as

Q-141   Stump Cross: charcoal   4550 ± 310 b.c.

This 'Atlantic' date, somewhat towards the end of the known Mesolithic age range, nevertheless is entirely agreeable and fits the pollen evidence from Verona Conway's 1947 survey, suggesting that much of the south

Pennine blanket peat had originated early in the Atlantic period.

Research continued in the twenty-odd years since Stump Cross was published has shewn that sites of the microlith users are plentiful and widespread through the southern Pennines in an altitudinal belt between 350 and 500 m OD, and that they represent Mesolithic hunting groups who, in all probability, made use of recurrent burning of the more or less open woodlands of the region as a means of fostering and controlling the game population on which they lived. Radiocarbon dates suggest that exploitation of this kind possibly extended through as much as 4000 years and lasted into the Middle Bronze Age. Certainly carbon particles are everywhere abundant in the upper mineral soil and it has been strongly suggested that the forest destruction by burning probably at least hastened replacement of woodland by the blanket bog under the increasing oceanicity of the Atlantic period. Likewise Frank Mitchell has suggested that in Ireland actual cultivation of the soil by promoting leaching and pan-formation has led to the initiation of blanket bog growth from the onset of forest clearance and occupation in Neolithic and later time. A considerable spread of dates for the onset of bog formation no doubt reflects the strong influence exerted by the local factors of altitude, slope, local hydrology and rock formation as well as human forest clearance and husbandry. This variability in age is encountered whether one judges by the pollen diagrams or employs carbon dating. It naturally extends to the dates found for the tree-remains discovered so frequently in the basal layers of the blanket bog.

Fig. 74. Pollen series recovered by Dr D. Walker in 1956 from the Mesolithic site under blanket bog in the southern Pennines at Stump Cross near Grassington at an altitude of about 1200 ft (390 m). It was evident that human occupation occurred early in zone VIIa, i.e. the early part of the Atlantic climatic period, at which time the blanket bogs also began to form. This was the earliest clear demonstration of the age of a Pennine microlithic industry and was subsequently borne out by radiocarbon dating.

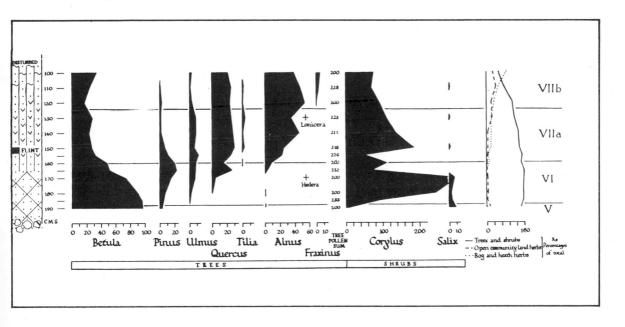

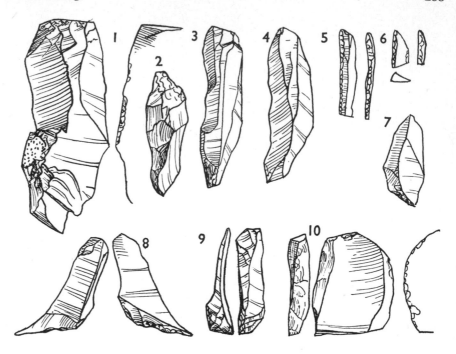

Fig. 75. Artefacts of flint and chert from the site at Stump Cross where a Mesolithic occupation was identified in the lowest layers of the blanket bog and was referred to pollen zone VIIa (see Fig. 74).

Tree stubs and wood-layers

As with the 'bog-oaks' of the East Anglian Fenland the sight of massive trees emerging from below a cover of deep peat has never ceased to excite speculation. This holds no less for the layers of pine stumps beneath the blanket bog of north-western Scotland, Galloway or the Cairngorm, and the buried wood-layers high on the Pennine slopes, than for tree-layers exposed in cutting through such raised bogs as those at Borth or Shapwick Heath. How often, through the years, have I been presented by a returning traveller with a piece of wood from such a buried giant seen exposed under deep peat with the hopeful request that I kindly inform the finder of its age! It will already be apparent that any meaningful answer had to involve field work on the local bog stratigraphy and comparative pollen analyses which, in remote regions, might themselves yield only the sparsest information. The advent of radiocarbon dating admittedly made it feasible to date the actual wood-sample, but it was doubtfully worthwhile to employ so expensive a facility without first becoming familiar with the detailed context from which the wood-sample was taken.

The fact that the fossil timber of the peat bogs tends often to occur in discrete horizons, the so-called 'buried forests' or 'wood-layers', is so evident that it is seldom commented upon. It arises of course from the fact that the fossil wood is the product of two contrasting sets of environmental conditions, firstly those of good drainage and at least moderately good nutritional conditions required for robust tree growth, and secondly, following upon this, contrasting conditions which kill the trees and preserve their remains, at least in part, before they are removed by decay. The second stage is apparently always that of the swift water-logging of the forest floor, when deoxygenation kills the tree roots whilst growth of wet acidic and mineral-deficient peat offers no opportunity for tree regeneration. There can be no doubt that in general the buried tree-layer has its origin in this overturn of ecological conditions whether local or general. This is true to whichever of the main categories of causation the buried forest may belong. The first of these is the progress of an autogenic vegetational succession, such as we have described in the stratigraphy at the base of the coastal raised bogs, where the sedge-fens, gaining height and becoming less wet, are invaded by fen-woods whose character progressively changes from wet alder–oak communities to intermediate oak–pine woods beneath which is established the carpet of *Sphagnum* moss that eventually kills and entombs the woodland. So far as the geographical and drainage relationships of the area impose widespread uniformity of vegetational change, the buried layer will tend to be of the same age: this will naturally be emphasised if a general turn of climate towards dryness should reinforce and speed-up the drying of the fen surfaces during

Plate 65. Fossil pine-stools exposed on the shore of Clatteringshaws Loch, Kirkcudbright, by a recent fall in water-level. Erosion has exposed the tree layer that underlies the adjacent raised bogs. These stubs have a radiocarbon age of 3130 b.c.

their tree colonisation and clothing by fen-woods. Where the extended lateral growth of raised bog with time leads to water-logging of the adjacent mineral soil, we shall again find a seral change. As marginal water-logging invades the standing forest and kills and then entombs its trees, here the resultant fossil tree-layer will exhibit a gradient of age although no doubt tending to uniformity if increased climatic wetness should have operated to accelerate the marginal bog growth. A very similar process, often rather grandly referred to as 'paludification', results in the burial and preservation of the tree-layers that so commonly lie upon or close above the mineral soil now found below vast stretches of blanket bog. Here there is little scope for attribution of water-logging to access of drainage water from without and the overturn of ecological conditions will generally have been the consequence of a climatic shift towards increased oceanicity, possibly enhanced, as we have already seen, by such practices as burning and grazing of the pre-existing woodlands. The vegetational succession induced in this way is evidently less autogenic (self-induced) in character, and has at least a large allogenic element in its causation, that is, of external control.

Thus far we have considered only the basal forest layers, but very often in both raised bogs and blanket bogs particular horizons in the middle and upper strata of the peat also display notable concentrations of the stumps and fallen trunks of trees that evidently grew profusely on the bog surface before they were killed by the worsening conditions. There can be little doubt that such forest layers as these were a response to a temporary climatic shift inducing drying-out of the bog surfaces, and indeed tree colonisation, usually by pine and birch (and their subsequent destruction) very often accompanies the other stratigraphic indications of the recurrence surfaces of whatever age these may be.

In the Scottish blanket bogs the tree-layers occurred with such consistency that they were designated as representing two periods, the 'Lower Forestian' and the 'Upper Forestian', separated by phases referred to as 'Turbarian'. As it came to be accepted that these phenomena were all Flandrian, i.e. Post-glacial, in age, attempts were naturally made to link the two forest periods with the two drier, more continental of the Blytt and Sernander climatic periods, but where it ultimately has become possible to apply radiocarbon dating to carefully collected and pollen-analysed samples, this simple hypothesis has had to be abandoned. Nevertheless the results of Dr Hilary Birks and Dr Roy Switsur by 1975 had shown that the Scottish pine wood-layers were clearly in two age groups, the one broadly between 4500 and 4000 years B.P. and the other less sharply defined, between 7000 and 6000 years old. In north-western Scotland only the younger 'forest-layer' has been found, and in Galloway only the older, whereas forest-layers of both ages are represented in the Cairngorm region, still the centre of the native forests of 'Scot's fir'. There seems little doubt that climatic change has been ultimately responsible for the presence of the two forest-layers, especially as very similarly dated layers occur in western Sweden, as also in Ireland, and one cannot but remark that the 'Upper Forestian' apparently corresponds with the earliest of Granlund's five recurrence surfaces, RY I (c. 2300 B.C.)

In our preoccupation with buried trees in relation to bog stratigraphy we should not lose sight of the long-recognised possibility that an important registration of past climate may be offered by the presence of fossil trees *in situ* at altitudes above the present tree-line. This has indeed long given particular interest to reports of the tree-layers below and within the upland peats and we note that Moss in his *Vegetation of the Peak District* (1912) suggested that trees formerly grew at altitudes 250 ft (76 m) higher than at the present time. More precise information comes from the National Nature Reserve at Moor House to the east of Cross Fell where a basal forest-layer, mainly of birch, but also of willow and juniper, extends up to the 2500 ft (760 m) contour, in which vicinity it is 'often absent'. This is contrasted with a present tree-line at about 1750 ft (530 m) in the South Tyne valley. It is difficult not to see this contrast as representing a real climatic change since the fossil forest grew about the time of the Boreal/Atlantic transition, and the more readily since this is generally held to fall within the Flandrian thermal maximum, or hypsithermal stage of the Post-glacial as certified by a big range of biological and geological evidence. Again and again throughout upland Britain, bog profiles reveal buried timber evidently far higher than the existing tree-line, and the argument holds good whether we are dealing with the hairy birch (*Betula pubescens*) as at Moor House and most of the Pennines, pine (*Pinus sylvestris*) as so commonly in Scotland, or oak, hazel, rowan and alder that are also commonly recorded.

Unfortunately it is a matter of extreme difficulty to determine at the present day the *natural* position of the tree-line, for not only does it differ greatly with aspect and slope, but it is liable to great depression by the grazing, notably by sheep, now so prevalent in mountain regions, and by the burning that has often been undertaken in furtherance of this pasturage. The observant naturalist will have seen those small isolated stands of dwarf trees that occupy an occasional craggy slope unattainable by sheep and fire; these are often one's only indication of where *natural* tree growth might occur in a landscape largely unaffected by man and his animals. We may ask ourselves moreover, if it is so hard to determine the *tree*-line, how much harder it will be to fix the natural upper limit of *forest*, i.e. the present-day equivalent of the buried forests beneath the blanket bogs.

Bog erosion

The concept that blanket bog is a natural climax requires that it should terminate a natural vegetational succession in a given climate, and that once established, it should hold its dominance by continuous regeneration so as not to be displaced by any other vegetation type. Although there is a great deal to be said for following Osvald and Tansley in regarding the British blanket bogs as a vegetational climax, it might, at first sight anyhow, seem to be contradicted by the widespread and advanced erosion seen in such bogs all over the British Isles. Quite apart from the special phenomenon of the bog-burst already mentioned, whenever one visits the upland peat lands, one finds the same desolate and indeed eerie scene of progressive destruction where confluent gullies dissect the surface, exposing the underlying peat to drought, frost, rain and wind action that swiftly eat back the channel walls, widening and deepening them until the mineral soil beneath has been reached. Between such gullies the peat comes to stand as blackish-brown hags, increasingly isolated from one another and supporting vegetation less and less like the original cover and characterised by increasing abundance of the few species tolerant of the increased drainage, especially cotton-grass (*Eriophorum vaginatum*), crowberry

Plate *66*. Holme Moss, southern Pennines. Blanket bog at 1700 ft (520 m) shewing very advanced and extensive peat erosion. The residual peat surface where intact is clad with crowberry (*Empetrum*). (1960)

(*Empetrum nigrum*) and cloudberry (*Rubus chamaemorus*). In some instances large stretches of mountain slope or summit have been almost
wholly denuded of bog, and it is one's first impression that the residual
bogs no longer have the power of healing the vegetational carpet by
renewed growth. However we have earlier mentioned evidence of
repeated regeneration on Dartmoor, and the experience of the staff of
the Moor House Nature Reserve in Upper Teesdale suggests that
regrowth there is common. It may be that the paucity of examples of
regeneration is evidence only of the suddenness and severity with
which the erosion has been initiated, and this chimes with the general
view that severe peat erosion is a relatively recent phenomenon,
perhaps of the last two or three centuries, whilst the fact that it is a
phenomenon to be seen so widely also indicates that it may well have
been the consequence of some generally operating outside cause.

It is certainly tempting to pursue this view by invoking a widespread
climatic change as responsible, but many difficulties stand in the way of
this kind of explanation. For instance there is the lack of evidence in the
bog stratigraphy where, over several thousand years, it seems that
phases of dryness led to tree colonisation and widespread dominance of
cotton-grass and heather, but not to erosion. Nor equally does the
stratigraphy show evidence of increased wetness except by change of
the vegetation to a wetter facies: there is no evidence of previous
gullying and large-scale erosion.

To those who were primarily familiar with the southern Pennines,
especially in the first half of this century, it seemed natural to look to
industrial pollution of the atmosphere as possibly responsible for the
widespread peat erosion. Later on Verona Conway was able to shew
that in the peat bogs of the southern Pennines the upper 8 in (20 cm) or
so of peat contained a high proportion of black soot and she was able to
record substantial concentrations of sulphur dioxide in the rain falling in
the region of Sheffield. Although it is likely enough that the bog
vegetation has indeed been much modified by these effects of industrial
pollution, it has to be remembered that the widespread bog erosion is by
no means limited to the areas within range of these gifts of civilisation,
and is to be found equally on the high plateaux and mountains of the
north and west, and in regions alike of Ireland, Scotland and Wales
where industrial contamination cannot possibly be invoked.

Increasingly the opinion of the leading investigators has come to look
for the cause of bog erosion primarily in increased human interference
with, and misuse of, the bog surfaces, partly by grazing and more
substantially through repeated burning in the interest of grazing. It is
well known that some of the main hummock-building *Sphagna*, such as

Plate 67. Wide gullies exposing the mineral soil in eroding blanket bog near the Snake road, between Sheffield and Manchester: 1600 ft (490 m). (1960)

S. imbricatum, are very susceptible to burning and that this species for instance is now largely absent from bogs in which it composes much of the upper peat. An explanation along these lines agrees well with the latest estimates of Tallis that widespread bog erosion seems to have begun, in the southern Pennines at least, only some two hundred years ago, the date at which, incidentally, Dr Conway thought industrial pollution and take-over by cotton-grass had first occurred in the bogs near Sheffield. It must necessarily be by indirect means that one dates the beginnings of disappearance of peat layers already vanished two or more centuries ago, so that Dr Tallis was constrained to measure the rates of enlargement of gullies and the yearly load of peat collectable from the drainage streams in relation to the calculated volume of peat bog already eroded.

Thus far, rather little responsibility has been attached to the widespread practice of peat cutting, which automatically operates first on the more accessible bog slopes. Tallis has demonstrated how the drainage of the blanket peat is chiefly within the body of the looser upper peat and at its contact with the top of the much less permeable black lower peat. When gully erosion transects this horizon it hastens drainage along the line of the interface of the two peat types and this certainly also happens, often rather suddenly, with marginal peat cutting. An increase in peat extraction in a sensitive bog could certainly initiate or hasten a local degree of erosion, and as with burning and grazing it is an activity that might well have reflected sudden growth in population with the industrial revolution and a rising degree of land exploitation.

15

The archive appraised

Preservation of the evidence

The effectiveness of the acidic mires as archives of past events and cultures long forgotten, lies to a great extent in the strongly preservative qualities of their peat, which, by being water-logged and anaerobic and at the same time acidic, prevents the decay of organic material, thus securing objects of wood, wool, leather and hair against decay, although bone is quickly dissolved.

One of the earliest and most dramatic applications of pollen analysis to prehistoric archaeology was associated with the recovery of a large woven cloak from a raised bog in Västergötland, where it had been folded and laid down as a votive offering in a pit cut down from the bog surface. By comparison of the pollen diagram at this site with those where there had been found other artefacts of recognisable age, such as a fine gold torc for example, von Post established that the cloak belonged to the Late Bronze Age. Its wool was so perfectly preserved that the detailed manner of its weaving and wearing could be fully made out. By 1925 pollen analysis was thus evident as an important aid in deciphering of the bog archives.

Certainly no less startling has been the recovery from Danish peat bogs (though not from British ones) of a considerable series of bog burials of human beings, the most famous of which is the corpse of 'Tollund man'. His skin, nails, hair and fur-lined cap were meticulously intact, as was the plaited rope that strangled him and still encircled his neck. So well preserved indeed was his stomach that the fruits and seeds of the porridge he had last eaten were all recognisable, permitting the engaging Danish title to the publication of 'Tollund Mandens sidste Maaltid' (Tollund man's last meal): a notable pre-execution meal indeed!

Of course a large proportion of the recovered artefacts have been of wood. Thus the discovery of the wooden haft of a Neolithic axe in a Danish peat bog was a necessary step in reconstruction of the way in which polished stone axes were used, and preceded the experimental

Plate *68*. Tollund man. The head of the corpse found buried in the Danish peat bog in 1950, with the leather cap he was wearing. The picture conveys very well the extraordinary state of preservation of the organic material. See also Fig. 76. (Photograph by K. Thorvildsen, 1950.)

forest clearances for shifting cultivation of cereals made so successfully by Iversen and Troels-Smith (see Fig. 66).

I once saw in the National Museum of Ireland a long straight-handled wooden scoop recovered from a peat bog and of uncertain purpose. Later one met references, in accounts of mediaeval turbary rights, to just such an instrument, a 'dydal', employed to ladle out peat from the bottom of water-logged peat diggings. Further, one gathered that still at about 1930, the East Anglian River Yare was ceremonially navigated by the members of an ancient society of 'Dydallers' and that, from some convenient boat, they employed a dydal for clearing water-weed from the river channel.

It will naturally have become apparent that these preserved wooden objects, recovered in such numbers from the Somerset Levels, afford the most fortunate and effective links with existing and remote patterns of life. From the raised bogs of Somerset have come the yew-bows already mentioned, the Shapwick monoxylous boat and, in later years, a wooden 'God dolly' and a mallet associated with the driving of piles. Above all of course, they have yielded the remarkable evidence of wooden trackway construction, still being so actively extended for structures of the Neolithic as well as of the Late Bronze Age.

How far the Glastonbury Lake Village is to be reckoned as mire or as lake margin is hardly material, but the painstaking records of Bulleid and Gray encompass a vast range of wooden artefacts including apparently both pile-dwellings and crannogs, with flooring, walls, ladders, palisading and boats, together with pieces of weaving looms, ploughs, wheels, wooden rungs and all kinds of ejecta that landed in the lake muds.

It will of course be apparent that the bog archives naturally comprise all those plant-remains of which the bogs themselves are composed, remains that are still recognisable, and whose stratification informs us of the nature and course of the bog's growth. At the lower end of the size range are naturally the pollen grains, from both the bog surface vegetation and the whole region, that in such vast numbers constitute the means of pollen-analytic reconstruction of past vegetational history. Similarly, though I have made little reference to them, are the animal microorganisms, mostly of specialised groups tolerant of high acidity and low mineral status. The bones of all vertebrates quickly disappear in the acidic peat, although now and again one finds recognisable trace of large animals in such residues as the cluster of elk droppings found at the site of the dug-out boat at Shapwick Station. These too were of about the same age as the boat, i.e. the beginning of the Iron Age.

The organic plant-remains are very commonly in their position of growth, but often display the effects of compression of the peat. This is especially so where vertical stems or roots, like the unbranched black roots of giant sword sedge, have grown down into lower peat which has afterwards compacted. The roots will now be folded into sharp zig-zags, by unfolding which we can of course gauge the extent of the compres-

Fig. 76. The plaited leather noose found round the neck of Tollund man, the buried bog-corpse from Tollund, 1950. (Drawing by K. Thorvildsen, 1950.)

sion undergone. The wooden stockading piles of the Glastonbury Lake Village were seen by Bulleid to have been folded similarly, as he judged, no doubt correctly, by contraction of the organic muds after the stakes had softened by progressive decay. It seems hard to attribute sharp folding of this kind to the distortion of fresh timber under the impact of being driven down by a mallet, even into a soft clay sub-stratum.

The bog records give the evidence we should expect of preservation of objects of flint, jade or other stone and of pottery. Artefacts of metals are more variable in condition when recovered from the bogs. The facility with which soils are leached of iron by humic acid points to the general susceptibility of that metal, objects of which seldom survive, or at best only badly, in the acidic peat. On the other hand pewter and bronze are excellently preserved save that bronze may suffer severe corrosion at an oxidising interface, as could be seen clearly in the case of the bronze spearhead shown in Fig. 45, which had evidently rested for some time only partially embedded in the bog surface before being fully covered.

There is a further sense, beyond the burial of evidence, in which the acidic mires serve for the preservation of scientific evidence. This lies in their role, whilst they still remain at least partially active, as refugia for species of plants and animals elsewhere in danger of extinction. Such mires have often been, and continue to be, relatively free from the demands of 'reclamation' and persist as natural habitats in regions otherwise drastically altered. They offer, of course, by their wetness,

Plate 69. Butterwort, *Pinguicula vulgaris.* Because of special requirements for moisture, acidity and freedom from overshading, certain plants today find refuge in the peat bogs. Among them is the butterwort, an insectivorous plant that catches insects on the sticky glands of its yellowish-green leaves. (Photograph by W. H. Palmer.)

Plate 70. The arctic birch, *Betula nana*, that had a wide distribution in the British Isles at the end of the last glacial period. It now survives in a restricted northern area. Acidic peat bogs, which are part of its natural habitat, offer localities in which it has locally persisted to the present day, as for example on Widdibank Fell, upper Teesdale. (Photograph by W. H. Palmer.)

mineral deficiency and acidity, a very limiting habitat but this at least confers the advantages offered by restriction of competition to the specialised flora and fauna. Accordingly many species survive in the acidic bogs, often despite burning, peat digging and some drainage, to yield evidence of a former more extensive occurrence than they now enjoy. We have already cited the instance of several of the individual bog plants whose range in the south and east especially has become disjunct and vestigial.

Sometimes the bogs offer the special advantage of reinforcing the evidence of present distribution by the presence of fossil remains within the bog deposits. Thus at Widdybank Fell in Upper Teesdale there was found growing a small community of the dwarf or 'arctic' birch, *Betula nana*, far south of its native Scottish mountain localities. The question of whether this was a survivor from the end of the last glacial period was answered by the recognition, in the same peat bog, of the characteristic pollen of the dwarf birch through successive zones of the intervening Flandrian, and even a well-preserved leaf from the mid-Flandrian. This evidence decisively pointed to the Late-glacial origin of the famous Upper Teesdale flora, an issue of great relevance to the controversy that had to be considered by special committees of both Houses of Parliament, before construction of the Cow Green reservoir was authorised.

We have already mentioned the case of *Scheuchzeria palustris*, the bog plant whose papery rhizomes have been found in relict peat bogs in many parts of Britain, whilst the plant itself very barely survives in this country. It is not surprising that similar instances should be known from the mosses that so often find suitable habitats in the acid mires, and in 1941, during the excavation of the Huntspill Cut in the Somerset Levels, there was found well-preserved material of *Meesia tristicha*, a species now known living at only one (Irish) station in the British Isles but here attested as present in the middle Flandrian. Bogs elsewhere have subsequently yielded similar evidence of a former more extensive range and Dickson and Brown, from raised bog peat in East Anglia, identified another species of the same genus, *Meesia longiseta*, not now known surviving at all in the British Isles. These identifications conform to the more abundant evidence of present and past range on the European continent, that both species are glacial relicts.

Another quite different instance of the bog evidence supplying a factual basis for plant geography is that of the common beech tree (*Fagus sylvestris*). On the strength of a statement by Julius Caesar that he had found in Britain all the trees of Gaul 'except spruce and beech', botanists had long held the beech to be a post-Roman introduction. The growing volume of pollen analysis was casting doubt upon this view when one of the sharpened stakes recovered in 1942 from the Shapwick Heath trackway proved upon microscopic analysis to be of beech. It was accordingly beyond doubt already native in the Late Bronze Age, a conclusion later confirmed by a similar identification of a stake from the earlier, *Neolithic,* Blakeway Farm track. In all probability the words used

Plate 71. *Ledum palustre,* an evergreen ericoid shrub found right across northern Europe and Asia. Here it is pictured growing on Scaleby Moss where, however, as in one or two Scottish bogs, its native status is doubtful. The answer ought surely to be in the local bog archive! (1955)

by Caesar (*atque fagem*) and taken to refer to 'beech', should have been translated to indicate 'sweet-chestnut', a quite valid alternative, and one pointing to a tree for which there is *no* evidence of native status.

Advantages of the bog record

The remarkable preservative properties of the peat in the acid mires are but one of the many qualities that confer on such mires, most especially the raised bogs, their remarkable advantages as archives of past events. What we learned in the twenty-five years or so when we taught ourselves to read the events locked in the western peat deposits, convinced us of their extraordinary trustworthiness and wide relevance. The general absence of tree growth upon the bogs means that the bulk of tree-pollen incorporated in their growing surfaces represents the regional woodland, whilst that from the dwarf cover of acidicolous shrubs is easily recognised and segregated in counting. Thus the bogs, which remain generally as 'islands' or 'holes' in the prevalent mantle of forest, offer an ideal basis for the establishment of *regional* pollen zones. They have advantages over lake sediments not only in the rapidity of accumulation, which allows the zones to be spread out vertically, but in the relative absence of reworking of the receptive surface by wading cattle, by burrowing animals and the shifts of deposition and erosion as conditions change. The bogs moreover are free of the necessity of recognising, measuring and allowing for an intake of water-borne pollen which often constitutes a large proportion, and one of variable origin, in the pollen intake of lake sediments.

It is not now generally recalled, but in the early days of the technique, a good deal of effort was devoted to testing the possibility that significant amounts of pollen could descend from the bog surface to lower levels, but always the results proved negligible.

The extreme suitability of the bog deposits to pollen analysis was of course outstandingly important when the method offered the only broad means of chronological comparison across Britain and western Europe, and through a large part of the Flandrian period. Even after the widespread adoption of radiocarbon dating from about 1960, pollen analysis remains of the greatest importance, not only for correlation of deposits unsuitable for radiocarbon assay, but also as providing direct information on the development of disforestation and agriculture, and on the progression of bog evolution.

Apart from the indirect evidence of climatic conditions provided by their role in contributing to the pollen-analytic history of vegetation, the raised bogs provide registration of climatic shift more directly through their own stratigraphy, as we have seen in the many instances of tree-layers, flooding horizons and regeneration surfaces, the last of these categories illustrated above all by the earliest and most striking *Grenzhorizont*, the original recurrence surface to which Weber drew attention.

We have had the most convincing demonstration of the bogs as repositories of spaced-out archaeological material from the Mesolithic culture onwards. Past human societies made little direct use of the bogs themselves for settlements, save marginally, but the influence of contiguous settlements was often, as in the case of the wooden trackways, still powerful. Thus a continuing scatter of archaeological objects has littered the bogs, and, thanks to intensive hand-cutting of peat, has been fairly fully recovered. Moreover the waterways and pools so often associated with, and later overgrown by the bogs, were often the site of pile-dwellings or crannogs with their associated platforms, causeways and boats, now conveniently preserved and stratified. Now and again the flooding from upland hillsides denuded of trees by prehistoric man, is also registered as a clay wash stratified into a peat bog.

It is no accident that peat bogs should again provide such generous evidence of former changes in the relative levels of land and sea. As the great Post-glacial rise of ocean-level affected the big coastal estuaries of today, the infilling of tidal and then brackish-water clay provided extensive level flats highly favourable to the development of raised bogs, and there they remain, right round our western coasts, ideally situated to register, in an alternation of fresh-water acidic bog and brackish or estuarine clay, subsequent minor transgressions and regressions of the sea.

The adoption of radiocarbon dating gave added speed and certainty to exploiting all the archival virtues of the peat bogs, offering as it did, an absolute physically based dating system for all the phenomena that the bogs could reveal. What was so outstandingly favourable was that the peat, particularly of the raised bog, from top to bottom was directly suitable for assay, being plant material whose active carbon had been safely locked away as soon as the cellulose and lignin of the plant body was synthesised. There remained only the need to guard against the downwards transport in the bog of soluble humic material, and penetration to lower layers by descending rootlets. Neither constitutes a very serious problem, partly because of the very favourable circumstance that raised bogs often grow so rapidly that carbon dating samples can be very widely spaced. This in itself allows the analysis of quite short historical events, as Eric Willis and Alan Smith had proved in their investigation of forest clearance and regeneration at Fallahogy.

Postscript

I have constructed this book largely around my own involvement with the ombrogenous bogs, from my introduction to them in the magical Irish countryside in 1935, through to the rather arbitrary date about 1960 when circumstances had deflected me into other major activities. The interest in the rain-fed mires had properly begun with familiarising oneself with their morphology and ecology, indeed concern with them as active, growing entities. On the other hand this period coincided with the first development in this country of the pollen-analytic discipline, the notion of Quaternary research as a field of study had begun to crystallise and in 1938 I had addressed to the University of Cambridge a considered plea for the creation of an organisation suitable to promote it in the departments most concerned.

It became apparent through the next decade or so, what admirable sources of scientific and historical information were these mires, even when partially drained and derelict, and repeated excursions to the Somerset Levels were aimed on the one hand to take advantage of each

Plate 72. An international assembly of distinguished botanists and 'bog archivists' descending from an esker ridge near Clonmacnoise, in central Ireland, 1949. Those nearest the camera are (left to right) J. Iversen, R. Nordhagen, K. Jessen (back view), H. Osvald, F. Firbas. On the slope are G. Negri, R. Tüxen, E. Hultén and W. Lüdi.

fortunate and fortuitous find in the turbaries, and on the other to promote clarification of some particular aspect of the history of the region. In this manner our studies between 1935 and 1960 marched within a rapidly expanding progression of Quaternary study, pursued by increasing numbers of active research workers in this country and elsewhere. This study has of course greatly expanded since 1960 and is still extremely active, but it is no part of my present purpose to summarise the results of this period. I believe that by 1960 the main lines of investigation in this sector of Quaternary bog investigation had been well established, and by confining oneself to studies in which the personal interest has been direct, one hopes to have conveyed a sense of the urgency and excitement that productive research always involves.

Apart from the earliest field trips, I have not referred to the many congresses, visits and field excursions where I learned so much from colleagues in Sweden, Denmark, Holland and North America, nor to the many meetings of learned societies, especially the British Ecological Society, where one was introduced to fresh mires and fresh concepts by those best qualified to explain them. Nor, despite strong temptation, have I included a record of a prolonged visit to New Zealand in 1956–7, when, in the footsteps of Osvald and von Post, I visited antipodean peat bogs directly comparable in morphology and hydrology with those of north-western Europe although they were largely built of plants in totally different natural orders, and despite the surprising fact that the hummock-building role of the *Sphagna* had, to a large extent, been taken over by various dwarf flowering plants.

On both ecological and archival grounds the case for conservation of at least a representative series of the ombrogenous mires must be undeniable. Nor is there much time to be lost. The scale of modern exploitation with its high degree of mechanisation and extensive business organisation is a very different matter from piecemeal manual peat cutting and shallow ditches nibbling into the bog margins. Moreover preservation is a matter of substantial difficulty because a maintained high-water level is a first requisite and this can be secured only by conserving the whole hydrographic unit, the entire bog. Thus one is faced with the preservation of a structure possibly a square mile (2.5 km²) or more in extent. It is simply impracticable to attempt conservation by leaving a standing block of peat bog protruding from a landscape of exhausted turbaries; it will be drained and dried out within a few years and its components will have shrunk and broken up. Likewise its surface vegetation will quickly have altered, the indigenous flora and fauna alike lost to a sea of casual invaders no longer excluded by maintained wetness.

Plate 73. Knud Jessen, of the Danish Geological Survey and Professor of Botany in the University of Copenhagen, the great pioneer archivist of the Irish peat bogs, and scientist of the greatest distinction. How he resembles his Jutish predecessor (Plate 68)!

Plate 74. International Phytogeographic Excursion to Ireland, 1949. On the right the Irish organisers G. F. Mitchell (above) and D. A. Webb: they were later the respective professors of Geology and Botany in Trinity College, Dublin. The two other members clearly shewn are F. Firbas (Göttingen), author of the great pioneer works on vegetational history of central Europe, and E. Hultén (Stockholm), outstanding authority on the plant-geography of the circumglobal north.

How grateful we should then be to the Nature Conservancy and its modern successor, the Nature Conservancy Council, for having gone so far, through purchase and management agreements, to secure a high degree of protection for many of the most highly regarded areas, and how grateful to the many local preservation societies, land owners and sympathetic peat firms who appreciate the needs of conservation of the mires and give active assistance to this end.

References

Bulleid, A. (1906). Prehistoric boat found at Shapwick. *Proc. Somersetsh. Archaeol. Nat. Hist. Soc.* **51,** 51.

Bulleid, A. (1933). Ancient trackway in Meare Heath, Somerset. *Proc. Somersetsh. Archaeol. Nat. Hist. Soc.* **59,** 19.

Bulleid, A. and Gray, H. St George (1911). *The Glastonbury Lake Village,* vol. 1. Taunton Castle.

Clapham, A. R. and Godwin, H. (1948). Studies of the Post-glacial history of British vegetation. VIII: Swamping surfaces in peats of the Somerset Levels. IX: Prehistoric trackways in the Somerset Levels. *Phil. Trans. R. Soc.* **233B,** 233.

Clark, J. G. D. (1963). Neolithic bows from Somerset, England, and the prehistory of archery in north-western Europe. *Proc. Prehist. Soc.* **29,** 50.

Clark, J. G. D. and Godwin, H. (1962). Neolithic long-bows of 4500 years ago, found in the Somersetshire peat. *Illustrated London News* 219.

Dewar, H. S. L. and Godwin, H. (1963). Archaeological discoveries in the raised bogs of the Somerset Levels, England. *Proc. Prehist. Soc.* **29,** 17.

Godwin, H. (1934). Pollen analysis. An outline of the problems and potentialities of the method. *New Phytol.* **33,** 278.

Godwin, H. (1940). A Boreal transgression of the sea in Swansea Bay. *New Phytol.* **39,** 308.

Godwin, H. (1941). Studies of the Post-glacial history of British vegetation. VI. Correlations in the Somerset Levels. *New Phytol.* **40,** 108.

Godwin, H. (1943). Coastal peat beds of the British Isles and North Sea. *J. Ecol.* **31,** 199.

Godwin, H. (1946). The relationship of bog stratigraphy to climatic change and archaeology. *Proc. Prehist. Soc.* **12,** 1.

Godwin, H. (1948). Studies of the Post-glacial history of British vegetation. X. Correlations between climate, forest composition, prehistoric agriculture and peat stratigraphy in Sub-boreal and Sub-atlantic peats of the Somerset Levels. *Phil. Trans. R. Soc. Lond.* **233B,** 275.

Godwin, H. (1954). Recurrence-surfaces. *Danm. geol. Unders.* RII, **80,** 19.

Godwin, H. (1955a). Botanical and geological history of the Somerset Levels. *Advancement of Science,* **12,** 319.

Godwin, H. (1955b). Studies of the Post-glacial history of British vegetation. XIII. The Meare Pool region of the Somerset Levels. *Phil. Trans. R. Soc. Lond.* **239B,** 161.

Godwin, H. (1960). Prehistoric wooden trackways of the Somerset Levels: their construction, age and relation to climatic change. *Proc. Prehist. Soc.* **26,** 1.

Godwin, H. (1960b). Radiocarbon dating and Quaternary history in Britain. *Proc. R. Soc. Lond.* **153B,** 287.

Godwin, H. (1978). *Fenland: Its Ancient Past and Uncertain Future.* Cambridge University Press.

Godwin, H. and Conway, V. M. (1939). The ecology of a raised bog near Tregaron, Cardiganshire. *J. Ecol.* **27,** 313.

Godwin, H. and Mitchell, G. F. (1938). Stratigraphy and development of two raised bogs near Tregaron, Cardiganshire. *New Phytol.* **37,** 425.

Godwin, H. and Newton, L. (1938). The submerged forest at Borth and Ynyslas, Cardiganshire. *New Phytol.* **37,** 333.

Godwin, H., Walker, D. and Willis, E. H. (1957). Radiocarbon dating and Post-glacial vegetational history: Scaleby Moss. *Proc. R. Soc Lond.* **147B,** 352.

Hibbert, F. A., Switsur, V. R. and West, R. G. (1971). Radiocarbon dating of Flandrian pollen zones at Red Moss, Lancashire. *Proc. R. Soc. Lond.* **177B,** 161.

Iversen, J. (1941). Landnam i Danmarks Stenalder. *Danm geol. Unders.* RII, no. **66.**

Mitchell, G. F. (1976). *The Irish Landscape.* Collins.

Moss, C. E. (1907). Geographical distribution in Somerset: Bath and Bridgwater district. *R. Geol. Soc.*

Moss, C. E. (1913). *Vegetation of the Peak District.* Cambridge University Press.

Osvald, H. (1937). *Myrar och Myrodling.* Kooperativa Förbundets Bokförlag, Stockholm.

Osvald, H. (1949). Notes on the vegetation of British and Irish mosses. *Acta Phytogeographica Suecica,* Uppsala, **26.**

Pearsall, W. H. (1950). *Mountains and Moorlands.* New Naturalist Series. Collins.

Praeger, R. L. (1934). *The Botanist in Ireland.* Hodges, Figgis and Co., Dublin.

Roles, S. (1960). *Illustrations*, Part 2. *Flora of the British Isles* by A. R. Clapham, T. G. Tutin and E. F. Warburg. Cambridge University Press.

Smith, A. G. and Willis, E. H (1961/2). Radiocarbon dating of the Fallahogy Landnam phase. *Ulster J. Archaeol.* **24/5.**

Tansley, A. G. (1939). *The British Islands and their Vegetation.* Cambridge University Press.

Turner, J. (1962). The *Tilia* decline: an anthropogenic interpretation. *New Phytol.* **61,** 328.

Turner, J. (1964). The anthropogenic factor in vegetational history. I. Tregaron and Whixall Mosses. *New Phytol.* **63,** 73.

Walker, D. (1956). A site at Stump Cross, near Grassington, Yorkshire, and the age of the Pennine microlithic industry. *Proc. Prehist. Soc.* **22,** 23.

Short glossary

allogenic (of vegetational succession) motivated by outside factors

autogenic (see above) motivated by the plant communities themselves changing their environment

calcifuge intolerant of lime

eustatic changes in absolute height of the ocean

eutrophic situations or communities associated with rich feeding (high concentrations of soil nutrients)

hydrosere the autogenic succession of vegetation from open water to terrestrial communities (*adj.* hydroseral)

isohyet line connecting places of the same rainfall

isostatic compensatory movements in the earth's crust resulting in differential changes in land- and sea-level

mesotrophic a moderate level or requirement of soil nutrients

oligotrophic meagre feeding: plants or situations associated with low levels of nutrition

ombrogenic (ombrogenous) (of peat bogs) rain-fed: essentially dependent upon direct precipitation

ombrotrophic plants or communities associated with rain-fed substrata: poor in nutrients

oxylophytes plants of acid soils

palaeoecology ecology of former plant communities and organisms

sere a vegetational succession: an ordered progression of plant communities (*adj.* seral)

soligenous (of peat bogs) fed by inflow of drainage water

topogenous (of peat bogs) landscape-dependent in origin: contrasting with 'ombrogenous'

Index

Bold type indicates where a topic is given special consideration: **GL** indicates mention in the Glossary.

Text figures are indicated by a page number and asterisk: the figure numbers are not given. *Passim* signifies 'scattered throughout'. Photographs are referred to at the end of each entry by their individual plate numbers: thus Pls. *21*, *32* . . .

English plant names have usually been given their Latin equivalents in brackets: indexed Latin names, however, are not accompanied by their English equivalents. Both Latin and English name entries carry the same page references. In general plant names in the text figures, legends and tables have not been included.

The positions of all the more important sites and waterways are shewn in the text figures cited in the index; grid-references, however, are given for many of the less familiar localities referred to, especially those in the Somerset Levels.